ETHICS
IN
SOCIAL
AND
BEHAVIORAL
RESEARCH

Edward Diener and
Rick Crandall

ETHICS
IN
SOCIAL
AND
BEHAVIORAL
RESEARCH

THE UNIVERSITY OF CHICAGO PRESS
Chicago and London

EDWARD DIENER is assistant professor of psychology at the University of Illinois at Champaign. RICK CRANDALL is associate professor at the Institute of Behavioral Research, Texas Christian University. Both have published a number of scholarly articles.

THE UNIVERSITY OF CHICAGO PRESS, CHICAGO 60637
THE UNIVERSITY OF CHICAGO PRESS, LTD., LONDON

© 1978 by The University of Chicago
All rights reserved. Published 1978
Printed in the United States of America
82 81 80 79 78   54321

Library of Congress Cataloging in Publication Data

Diener, Edward, 1946–
  Ethics in social and behavioral research.

Bibliography: p.
  Includes index.
  1. Social science research—United States.
2. Ethical problems.   I. Crandal, Rick, joint author.
II. Title.
H62.D53   174 .9 301   78-8881
ISBN (cloth) 0-226-14823-8

Contents

Preface

This book deals with the ethics of conducting scientific research and with questions about the role personal values play in research. As the scope and the success of the social sciences[1] have grown, there has been a heightened concern with such questions and an increasing realization that if scientific achievements are to benefit humanity, researchers need more than technical proficiency. Most of this book discusses how to insure the welfare of the individuals and communities studied in social science research. The chapters on social science and values discuss the role values play in social science research and the influence it has on society. The relationships between ethics, values, and research are among the most important problems faced by social scientists. As working researchers, we found no book that offered a detailed overview and broad introduction to this area. We hope that this one will be useful to researchers concerned with these issues and will fill an important gap in the eduation of present and future social scientists.

Education in the social sciences currently emphasizes technical proficiency. We think this stress is beneficial because we believe that research and the empirical perspective are absolutely necessary to scientific progress and human welfare (see also Haywood 1977; McGuire 1967). But we believe technical progress must be accompanied by a sensitive concern for values and ethics. A noted scientist, in stating that "science has taught us to be Gods before we have learned to be men," was suggesting that our technology has outpaced our wisdom and humanistic ideals. We believe that

1. The terms "behavioral science" and "social science" are used interchangeably in this book to refer broadly to research in the areas of psychological, social, and cultural processes. Both terms frequently connote a more limited scope than we desire, but unfortunately there are no other terms that denote the broad area of research dealing with human functioning outside the biomedical area. We do not mean to employ the terms behavioral and social science in the ideological or metatheoretical way they are often used.

researchers who are knowledgeable about and sensitive to issues of ethics and values are the best guarantee against abuses of science. We believe that responsible social scientists must maintain a balance between scientific proficiency and ethical sensitivity in order to prevent the misuse of science.

A major theme of this book is the idea that educated and concerned researchers provide the best safeguard against scientific abuses. In this spirit we have emphasized discussion of the pros and cons of important issues, reviewing their ethical considerations, and, we hope, stimulating readers to weigh them carefully. Rarely do we give ready-made formulas for making difficult decisions. In several places we have felt compelled to offer strict guidelines, but usually specific decisions are best made after thoughtful analysis by the individual researcher. Our approach may be too liberal for ethical conservatives and too constraining for laissez-faire investigators. As working researchers we have attempted to offer workable guidelines, not inflexible moralistic commandments. We have tried to achieve a balance by presenting the issues fully and offering a few guidelines, then encouraging educated scientists to exercise their good judgment. Our major hope is that the content of each section will serve to heighten the knowledge and awareness of these ethical and value issues for practicing and future researchers.

The book is designed to be relevant to a variety of social scientific fields such as anthropology, education, psychology, and sociology. Professionals in related areas such as industrial research, leisure research, and political science should also find the book pertinent to their ethical concerns. We hope to provide an organized overview of ethical issues and new insights for experienced researchers, but we have also attempted to present the material clearly enough and with adequate examples so that it can be understood by students who are preparing for or considering research careers. We have avoided discussing every ethical issue exhaustively, both so that the book may be read more quickly and because we believe that in practice a keen sensitivity to major issues is more desirable than full knowledge about everything that has been written. However, reference lists and recommendations for further reading are supplied for those who desire to pursue specific topics in more depth.

In an effort to avoid seeming sexist, we used the expression "he or she" throughout our original manuscript. But, because the material demanded so many personal pronouns, the publisher suggested using "he" for both sexes to avoid awkward constructions. We consented to this somewhat grudgingly, but we have tried to preserve a balance by making our hypothetical investigators both female and male.

In some cases we have constructed examples of ethically questionable studies in order to capture the essence of particular dilemmas. In other cases we have cited some actual research containing debatable or questionable ethical practices. Many discussions of ethics cite no specific studies in order to protect the reputations of the investigators. We have chosen to illustrate ethical points with real studies and have included specific references to them so that readers can examine the original studies in detail and make their own ethical judgments. If we did not identify the authors of studies, many readers would recognize them anyway, yet other readers would be prevented from making their own informed judgments for lack of a reference. Our use of specific examples is not intended to point the finger at individuals, and in no case do we mean to imply that an investigator has willfully acted unethically. Among the problematic examples are several of our own studies. Ethical concerns have only recently been widely discussed, and many of the studies we cite were completed in the 1950s and 1960s before such concerns and knowledge became commonplace. It may therefore be unfair to fault investigators personally for studies conducted when ethical standards were more poorly defined and less stringent and when ethical knowledge was less widespread. In addition, just because a study had *debatable* ethics does not mean that it was unethical—men of good will may differ on propriety. The examples and indeed the entire book are directed at heightening ethical sensitivity and knowledge.

This book evolved over a period of several years. Diener originally planned to edit a book of readings on research ethics, but it became clear that the necessary breadth of coverage could not be encompassed by the existing articles. Diener wrote the original drafts of all the chapters except chapter 10, and the final work is a result of discussions and revisions by both authors.

We would like to acknowledge the aid of those who have helped us. As psychologists we have benefited from the extensive work on ethics by the American Psychological Association. Naturally the work of colleagues in all disciplines has been critical in the development of the ideas presented here. We have cited examples and discussed issues from many areas and have tried not to be parochial. Several individuals have been very helpful. Felice Dublon made useful comments and spent innumerable hours in the library collecting articles and checking references. The book would have been impossible without the excellent typing skills of Genevieve Baker, Eileen Place, and Debie Whelchel. Their patience and fortitude were remarkable as they typed draft after draft of the manuscript. We thank Larry Wrightsman for his encouragement and comments, Phil Brickman for his helpful comments and for obtaining reviews, and S. B. Sells for comments on chapter 12, as well as other anonymous reviewers. We thank the many student reviewers for their comments on the early chapters. We also thank Cindy C. Haszier and Carolynn Crandall for major help on the index.

The excerpts from the codes of ethics of the American Psychological Association and the American Sociological Association are reproduced by courtesy of those organizations. The ethical guidelines of the American Anthropological Association were undergoing revision while this book was in press, and the new version was not available in time to be included.

1 Introduction and Overview

Psychological knowledge can bring man increased certitude, dignity and joy, and it can also enslave him. These antinomies are among the most exciting and demanding developments of our time. They have within them the seeds of ultimate tragedy or triumph. The stakes seem to be getting ever higher, and the rules of the game, embodied in ethics, ever more important.

Nicholas Hobbs (1959)

Absolute rules do not offer useful solutions to conflicts in values. What is needed is wisdom and restraint, compromise and tolerance, and as wholesome a respect for the dignity of the individual as the respect accorded the dignity of science.

Oscar Ruebhausen and Orville Brim (1966)

Why Ethics?

Ethical principles are vital to active researchers. They help social scientists achieve their values in research, avoid strategies that might endanger these values, and aid in balancing values that are in conflict. Ethical guidelines help insure that our research is directed toward worthwhile goals and that the welfare of research participants is protected. In other words, ethics are important because they help prevent abuses and serve to delineate responsibilities.

World War II sensitized both scientists and laymen to the importance of ethics because of the Nazi concentration camp experiments and the atomic bomb. Before that time it had been widely assumed that science was value-free and would automatically serve human welfare, and that scientists would invariably perform their research ethically. But during the war Nazi scientists conducted grossly inhumane experiments. For example, they immersed people in ice water and recorded how long it took them to die. In addition, the atomic bombs dropped on Hiroshima and Nagasaki and the subsequent nuclear arms race made it clear that the

1

products of science could be harmful. It has become increasingly clear that science will not invariably be beneficial unless wisdom and concern for others are combined with scientific research. As Julius Seeman (1969, p. 1028) said, "The existence of Hiroshima in man's history demonstrates that knowledge alone is not enough, and that the question 'Knowledge for what?' must still be asked. If knowledge in psychology is won at the cost of some essential humanness in one person's relationship to another, perhaps the price is too high."

When many of us think of ethics and values, we imagine topics that are boring, repressive, or irrelevant. We associate ethics and values with sermons about morality and with endless philosophical debates. But research ethics should not be a set of repressive moral dictates or irrelevant scruples imposed on us by self-righteous busybodies. They should be a set of dynamic personal principles that appeal to vigorous and active scientists who face difficult real-world ethical decisions. We believe in the importance of research and are convinced that most behavioral and social science research is ethical and poses no threat to participants. However, there are many cases that suggest that more attention should be paid to ethical concerns.

In a study described in chapter 2, the physiological changes accompanying emotional states were investigated. To induce fear, participants were given an unexpected electrical shock that most subjects thought could kill them. Believing the experiment was about hypertension, they did not know they would be subjected to this terrifying experience. Was it ethical to expose subjects to such fear? Is the informed consent of each participant necessary in such studies?

In a field investigation described briefly in chapter 5, Albert Reiss, a sociologist, studied police treatment of citizens. He was especially concerned with the nature and extent of police brutality. Reiss informed the police chiefs about the true nature of the study, but individual officers were misled about what was being observed. Although the officers were not individually given a choice about whether or not to participate, was the consent of the chief and the safeguarding of the officers' anonymity sufficient protection of the officers' rights?

In a series of negative income tax experiments funded by the United States government (see chapter 8), poor people in specific cities were randomly chosen to receive varying amounts of monetary assistance. Was the research unfair to

poor persons who were not given an opportunity to participate? Who has the right to decide the goals of such a study
that will strongly affect the lives of so many?

A scientist conducts a study to test her theory of interpersonal
conflict. A large set of data is collected and analyzed in many
different ways. Part of the data, when analyzed in one particular way, support the researcher's hypothesis. She reports
only the supportive data and analysis and does not mention
the nonsupportive findings (see chapter 9). In addition, a
graduate student who collected all the data is not listed as an
author of the paper. Was the researcher ethically obliged to
credit the assistant? (chapter 10).

Rainwater and Pittman (chapter 6) studied a large government-funded housing project in Saint Louis. Owing to the
nature of the study, they found it impossible to protect the
anonymity of some of the officials involved. Because of the
public concern and controversy surrounding the housing
project, the researchers' findings were likely to influence public
policies governing the project. To what extent were Rainwater
and Pittman responsible for the uses made of their findings?

These cases illustrate that important ethical issues arise frequently during research and that investigators must have a working
set of principles to guide their decisions. These cases also demonstrate the complexity of many of the ethical choices that must be
made. Behavioral scientists must understand ethical problems and
know about the available safeguards in order to make these moral
decisions competently.

What Are Research Ethics?

Ethics are guidelines and principles that help us uphold our
values—to decide which goals of research are most important and to
reconcile values and goals that are in conflict. Ethical guides are not
simply prohibitions; they also support our positive responsibilities.
For example, scientists have an obligation to advance knowledge
through research. They also have a responsibility to conduct
research as competently as they can and to communicate their
findings accurately to other scientists.

Many codes of morals and ethics exist, and they do not always
agree in their prescriptions of what is "good" and what is "bad."
This disagreement brings us to an important distinction between

different types of ethics: *wisdom ethics, content ethics,* and *ethical decisions.* These three types of ethics serve different functions and guide our actions in different ways.

Wisdom ethics (Garrett 1968) are the expressions of the ideal practices of a profession, but they are often not realized in concrete situations because of human limitations and competing values. For example, it would be an ideal ethic to assert that no research subject should ever be harmed by participating in a study, but it is usually impossible to guarantee this in practice. A participant may be inadvertently harmed in some unforeseen way despite existing safeguards. In other instances, subjects may be asked to voluntarily assume an unavoidable risk of harm in order to achieve some important goal that could not otherwise be reached. Even though wisdom ethics can rarely be achieved in practice, they do help scientists keep their goals and values firmly in mind.

Content ethics are explicit rules that state which acts are right and which are wrong. In the social sciences there are some generally accepted content ethics—standards on which virtually all scientists agree. For example, a universally accepted ethical rule states that a scientist must never harm participants unnecessarily. However, there are other rules or content ethics that are shared by some scientists but not universally accepted. For example, some believe that any use of deception by social scientists should be prohibited. Yet many other researchers would not accept this rule. Most content ethics are formally written down and published in the codes of the various scientific associations.

Although this book does describe some wisdom and content ethics, the major emphasis is on a third meaning of ethics—*ethical decisions.* The ethical or moral scientist makes individual judgments about research practices in light of his own values. According to this approach to ethics, the moral person is not one who blindly follows ethical codes, no matter how enlightened. The ethical decision-maker is one who realizes that his choices are related to values and weighs these values carefully when making important decisions. For the moral person there may be a few moral absolutes (Szasz 1967), but he realizes that most moral decisions must be made individually in each case (Smith 1969). This meaning of ethics emphasizes the process by which decisions are made as well as the final choice. The decision is made by a person who is educated about ethical guidelines, carefully examines moral alternatives, exercises judgment in each situation, and accepts responsibility for his choices. This book is dedicated to aiding the ethical scientist in such decisions.

Overview of the Problem

Ethical questions in the social sciences can be divided into three major areas: treatment of research participants, professional issues, and the relation between science, values, and society. Each of these areas is treated in a major section of this book.

TREATMENT OF SUBJECTS

As shocking stories about abuse of research participants have made the news, the treatment of subjects[1] has become a major area of concern in the biomedical sciences. In one unfortunate case it was discovered that live cancer cells had been injected into elderly patients without their knowing the contents of the injections (Edsall 1969; Lear 1966). In another case prisoners were experimentally infected with syphilis, and after years without treatment some of the subjects died of the disease. Indeed, a recent review of medical research suggests that in recent decades hundreds of ethically questionable studies have been conducted on medical patients (Beecher 1966a, b). Fortunately, abuse of subjects within the social sciences has been infrequent. Most social science research has been ethical and relatively harmless (Gergen 1973; Phillips 1976; Reynolds 1972; Steiner 1972). Unfortunately, there have also been cases of possible abuse of subjects within the social sciences, as the following examples will demonstrate.

Imagine yourself as a young army recruit and the unknowing subject of a psychological field study. You find yourself in a disabled DC-3 military aircraft, a United States soldier about to lose your life in the service of Uncle Sam. In your terrified state you are asked to fill out several forms before the airplane crash-lands. One form is a complicated questionnaire asking about the disposition of your earthly possessions in case of your death. Another form inquires into

---

1. There has been criticism of the use of the term "subject" for those participating in research because for some it connotes subservience and inferiority. "Participant" has been proposed as an alternative that connotes a more active role in the research enterprise. For the sake of variety we have used the terms participant, subject, and respondent interchangeably, and we do not imply a passive or inferior role by any of these terms.

your knowledge of emergency landing procedures, in order to furnish proof to the army's insurance company that proper emergency precautions were followed. After everyone completes his forms, the questionnaires are to be jettisoned in a metal container so that they will not be destroyed in the crash. But when everyone has finished the questionnaires, the plane lands safely, and only then do you learn that your intense fear of dying was unnecessary. The entire incident was part of a research study conducted by United States Army scientists helping the military gain information on how people react to stress (Berkun et al. 1962).

During the army stress studies other soldiers were subjected to seemingly even more frightening events, all in the name of science. During military maneuvers or "war games," recruits were left alone in an isolated spot. Artillery shells began to land all around them, apparently fired by gunners who did not know anyone was in that location. As the shells exploded progressively closer to the young soldiers, the psychologists observed their reactions to stress. In another simulation, recruits were led to believe that they were being bombarded by dangerous radioactive fallout. The ingenuity of the scientists was demonstrated even further in a simulated forest fire. The subjects were left alone in a small command post surrounded by forest. Through radio reports and a smoke machine concealed in the forest, they were led to believe that a huge forest fire was closing in on them from all sides. In the fifth simulation a soldier was led to believe that he was responsible for an explosion that had critically injured one of his comrades. The subject's task was to fix a broken radio so that help could be summoned from headquarters. While the unsuspecting soldier tried to repair the radio, he was continually reprimanded by his superior officer, who blamed him for the "serious injury." The studies revealed that this last situation, in which subjects believed they had caused serious injury to another soldier, produced the most extreme stress.

The army studies were conducted to obtain knowledge about stress and undoubtedly did serve a scientific function. Yet the participants in the research did not consent to undergo stress; in fact, they were not even informed that they were participating in a reseach study. They were deceived by elaborate hoaxes and placed under extreme stress. These experiments took place in the army, an environment in which a great deal of freedom routinely is lost. Therefore, some consider the studies less unethical than if they had been conducted with civilians. But similar practices, albeit usually less extreme, do occur occasionally in other social science research.

In part 1, the primary issues regarding proper treatment of research subjects are discussed in detail. Although outright mistreatment and serious harm have been rare in social science studies, there are still many reasons for studying ethics related to the treatment of participants. One reason is to eliminate the harmful studies that do occur. Another is to become aware of strategies that may enhance respect for subjects and their well-being without detracting from the research. And a third reason is to be better able to make decisions about what things may actually harm the individuals or groups being studied.

Four general problems that most often concern social scientists in their treatment of subjects are potential harm, informed consent, privacy, and deception. Chapter 2 deals with exposing participants to harm, risk, loss of rights, or stress. Examples of questionable research are described and the ethical criteria that may be applied to potentially harmful studies are discussed. Chapter 3 is concerned with informed consent—the degree to which participation is voluntary and the subject is knowledgeable about facts of the study that would seriously influence his decision to take part. Privacy and confidentiality are becoming important concerns in the modern world, and the questions involved when social scientists encroach upon these rights are the topic of chapter 4. Few techniques have raised more ethical eyebrows than research deception, and the arguments for and against it are presented in chapter 5. These four chapters should give the reader a knowledge of the basic ethical issues and safeguards in the treatment of subjects.

The issues covered in part 2 are also concerned with subject welfare but are more narrowly applicable to specific types of research. Particular ethical issues are encountered when scientists examine the everyday life of participants, especially when the research is conducted in other cultures or subcultures. This type of research, called by a variety of names such as cross-cultural research and ethnographic research, uses slightly different methods depending on the discipline. But common ethical problems are involved, such as unduly altering the culture under study, which are reviewed in chapter 6. A type of field research related to the issue of deception is the technique of disguised participant observation, in which the scientist secretly takes on some role in the subjects' everyday world to observe their behavior in natural settings. When social scientists misrepresent their identities during participant observation, the method raises serious ethical concerns that are reviewed in chapter 7. The last chapter in this section covers experiments in which a major goal is to change people in the "real world." Researchers implementing or evaluating new educational methods, attempts at clinical change, and community intervention all seek to create beneficial changes and to simultaneously measure the effects of their efforts. The issues that arise in change experiments, such as the problem of untreated control groups and the question of the unjustified imposition of our own values upon others are treated in chapter 8. Throughout part 2, in addition to the specific problems inherent in each type of study, the general issues of informed consent, privacy, and deception are applied to the research under discussion.

PROFESSIONAL ISSUES

Within each profession there are problems that directly influence its effectiveness but that are of only indirect concern to society as a whole. For example, ethics concerning publication practices are important to the dissemination of knowledge within a profession, but few citizens outside the discipline are concerned with these practices. Unfortunately, within many professions either there are no guidelines or there is an unwarranted assumption that everyone knows the guidelines. But when they do exist, the guidelines are often general, not widely publicized, and rarely taught to students. The young professional can thus come under fire for breach of ethics he did not know existed.

Although often only of parochial concern within the social sciences, these professional ethical concerns are sometimes extremely important. For instance, one such ethical dictum is both universal and absolute: a scientist must not falsify research findings. Yet unscrupulous researchers outside the social sciences have been exposed faking research results. In the well-known Piltdown hoax, a human skull and an ape jawbone were altered to simulate bones of primitive man. So discrepant were the Piltdown findings with the current theories of evolution that the ensuing controversy threw the science of physical anthropology into disarray. The hoax was so sophisticated that the debate over Piltdown man's place in evolution flourished for years and served to confuse theories of physical anthropology. Piltdown man was definitely proved a fake only after the perpetrator's death. More recently, William Summerlin of the renowned Sloan-Kettering Institute for Cancer Research confessed to an embarrassingly simple-minded fraud. He painted black patches on white mice and represented them as skin transplants from other animals. These cases illustrate that fraud does indeed occur in science and demonstrate that researchers may act in professionally unethical ways.

Once again the social sciences are fortunate that falsifying results does not appear to be common; but, alas, it does happen. For example, psychologist J. B. Rhine was the leader in parapsychological research for a quarter of a century. In his North Carolina laboratory, Rhine carried out carefully controlled research on extrasensory perception and other psychic phenomena. As Rhine grew older, he chose a brilliant young physician, Walter Levy, as director of the preeminent Institute for Parapsychology. Levy introduced computers and sophisticated automated devices into Rhine's laboratory so that data would be recorded mechanically. This automation was designed to circumvent errors and cheating by human data recorders. But then Levy himself was caught altering the data.

In one of Levy's experiments, rats had electrodes implanted in the "pleasure centers" of their brains and randomly received pleasurable stimulation. The research was designed to measure a psychokinetic effect: whether the rats could psychically influence the random stimulation generator and thereby receive greater numbers of pleasurable stimulations. The data were measured electronically and recorded by computer, supposedly untouched by human hands. But Dr. Levy's hands did get into the act. He would surreptitiously unplug the data counter at certain times (Asher 1974), making it appear that the rats had influenced the generator psychokinetically and thereby received more than a chance number of stimulations. Levy's deceit was discovered by his assistants, and the bright young scientist immediately resigned. Not only was his research career sadly ended, but suspicion was again cast on all parapsychological data in a field that had worked for years to gain acceptance within the skeptical scientific community. Recently, a case of possible fraud by a renowned British psychologist, Cyril Burt, has shocked the scientific community (see chap. 9).

There are also more subtle forms of abuse. "Covering up"

research data, concealing negative evidence, and incorrectly analyzing results are only a few examples of the distortion of truth by social scientists. These less blatant practices undoubtedly occur much more frequently than outright cheating and can undermine the scientific endeavors of social scientists.

Part 3 deals specifically with ethical relationships between scientists. Chapter 9 presents disturbing evidence of cheating and carelessness in data collection and analysis and reviews the importance of honesty and accuracy in scientific reporting. Chapter 10 is concerned with issues related to publishing and to grants and contracts. Chapter 11, on the subject pool, reviews the debate about the justifiability of using students as semivoluntary research subjects.

SCIENCE AND SOCIETY

The third major area of concern is the relationship between science and society. One aspect of this relationship is the degree to which societal concerns and cultural values should dictate the course of scientific inquiry and the questions asked—the value neutrality of science. Chapter 12 discusses the influences of cultural values on science. During the nineteenth century, the assumption that science was "value free" was widespread. It is now known that science cannot be totally value free and that all endeavors of man, including science, are influenced by values. Scientists often choose research questions based on personal interests and interpret data in ways influenced by their own beliefs. Since the social sciences cannot be totally value free or completely objective, important questions arise. To what extent can we as scientists be objective? In what ways should our deeply held values influence our research? An overview of these broad questions is given in chapter 12.

Whereas chapter 12 concerns the influence of values and beliefs upon science, chapter 13 discusses the complementary question: How will science and technology affect society, and who is responsible for guiding this process? The social sciences can also have a tremendous impact on society, even to the point of revolutionizing our conceptions of human nature, society, and culture. However, as the products of the physical sciences reveal, it is no longer an inevitable conclusion that this impact will be beneficial. To what extent are behavioral scientists responsible for the influence on

10

society of their knowledge and "products," and how can they exercise responsibility for the outcomes of science?

ETHICAL SUGGESTIONS AND APPENDIXES
At the end of the text is a short section containing our ethical suggestions for researchers. These are meant to be broad guidelines that will apply to most studies, and they do not attempt to summarize the material from individual chapters. Last are appendixes that contain further references and the ethical codes of several societies. As the following discussion indicates, the ethical codes of professional societies are only one of the types of guidelines available to the ethical investigator.

Enforcement of Ethical Codes
Regulations guiding social scientists exist at several levels. Legal statutes, codes of the professional societies, ethics review committees, and the personal ethics of the individual scientist are all major regulatory mechanisms. Laws have been passed in the United States to govern social science research, but thus far these laws are neither very restrictive nor very specific. For example, one major law that is enforced under the United States Department of Health, Education, and Welfare states that research funded by HEW involving human subjects must be reviewed by an institutional review board (Weinberger 1974, 1975). The law also prescribes that certain procedures such as informed consent be implemented when research entails risk to subjects. However, the law leaves much latitude to local ethics committees (called "institutional review boards") to make decisions in light of the specifics of each case. There is a fear among professionals of control by laws because laws tend to be unresponsive to changing needs. In addition, laws must apply to a wide variety of situations and thus cannot be sensitive to the specifics of particular cases. Since laws are largely created for scientists by another profession (politicians), researchers fear that statutory ethical standards will be less enlightened than guidelines created within the scientific community. As in all of life, self-control is preferable to control by others.

In addition to laws, most of the major professional associations of social scientists have created codes of ethics to help guide their members (see Appendix A for samples). These codes comprise wisdom gained from past research experience and the consensus of values within the profession. They aid the investigator because they delineate what is required and what is strictly forbidden. Ethics codes usually contain wisdom ethics and also point out ethical pitfalls, both of which are very helpful to the individual scientist. Ethical codes can often be more relevant to a scientist's problems than are general laws because they are written to cover the specific problems that are frequently encountered in the types of research conducted within that profession. In other words, codes sensitize the scientist to obligations and to problem areas within his discipline (Smith 1969) and indicate those areas where there is some consensus about good ethical practice. The ethical investigator should be educated about these formal codes in order to better formulate a personal, working set of ethical principles. Moreover, a scientist's personal code will mature as he acquires new experiences.

Should the researcher commit a serious and clear-cut transgression of his profession's ethical code, he is likely to be criticized and reprimanded. Although ostracism from the profession is a possible punishment for serious offenses, the mere threat of censure and strong individual consciences have usually been sufficient regulatory forces. In practice the professional codes serve as guideposts, rarely as absolute commandments. Ethical codes, like the law, evolve over time in response to changing research trends and increasing experience with new research techniques. Ethical codes simultaneously serve societal welfare and the self-interest of the profession.

Most institutions (e.g., universities) have research ethics boards, and sometimes there are additional committees within smaller units of the institution, such as colleges or departments. These committees make decisions about whether particular studies should be conducted as proposed, taking into account specific factors such as the abilities of the investigator, concerns of the local community, details of the research design, and the professional code of ethics of the scientist. Rarely do the institutional review boards reject behavioral science proposals out of hand, but they frequently recommend or require additional safeguards.

Without assurance from the institutional review board that the research is ethical, most federal funding agencies will not support the studies. Under HEW regulations, there must be a review mechanism and a set of guidelines at each institution receiving

HEW funds, and institutions have encountered legal and public relations problems when their review mechanism was inadequate (R. J. Smith 1977*a, b*). Institutional guidelines must be approved by HEW but may vary within wide boundaries. Often these guidelines require an ethical review of all research, funded or unfunded. However, what is defined as falling under the rubric of "research" varies between institutions. For example, observation of public behavior does not always require review. The guidelines also usually require specific safeguards if subjects are placed "at risk" in a study. This review mechanism leaves ethical control at the local level but assures that the large amount of research funded by the federal government will not be obviously unethical. Ethics committees are important because they provide more impartial judgments about ethical problems than most researchers can make about their own research and thus give the investigator additional perspectives on the issues. Because ethics committees are a primary source of outside advice, investigators should submit questionable ethical research to a committee or to concerned colleagues. Yet, though review boards can be a great aid, the individual scientist cannot transfer his own accountability to the committee but must continue to carry full individual responsibility for the research, regardless of committee approval.

The major regulatory force within society is individual self-control. People generally refrain from murder not because of fear of punishment or even respect for the law, but because most people in our culture believe murder is evil. Laws and codes serve as adjuncts to personal conscience as the major form of ethical control. Self-regulation should be especially effective among professionals because they have a high concern for societal welfare.

One reason individual judgment is crucial for decisions about research ethics is that the decisions often involve conflicting values that cannot be fully achieved simultaneously. For example, a scientist may value truth but also desire to protect the health of his subjects. If a research design requires so much stress that subjects' health is endangered, these two values may conflict. In this case the scientist must choose between them or achieve a compromise even before subjects are given a choice whether to participate. Often the most difficult ethical decisions are those between desirable but incompatible values. The moral scientist must carefully weigh the values involved and the potential alternatives, then personally accept

13

responsibility for the best human decision he can make (Smith 1969).

The personally ethical scientist, with the guidance of a professional code and local ethics committee, is the surest and wisest safeguard against ethical wrongdoing. Laws may be too rigid on the one hand and too easy to evade on the other. They may reduce the worst abuses, but they are unlikely to stimulate positively ethical research. An unethical investigator can usually devise ways to circumvent laws. Beecher (1966a, b) has made the point that the best safeguard is the *responsible* investigator who consults peers about difficult ethical decisions. If the individual scientist is thus the primary locus of ethical control, he must be educated and sensitized to critical issues. Professional societies cannot argue convincingly for individual scientific responsibility if their members do not have training, knowledge, and concern about ethical issues.

Conclusion

Ethics are not merely the results of a pretentious etiquette—they are expressions of our values and a guide to achieving them. Ethics are important to behavioral scientists because they help insure that our scientific endeavors are compatible with the goals and values we consider most important. The scientist who is committed to ethical research has a passionate dedication to truth and simultaneously seeks to protect the welfare of those studied. In addition, ethical scientists are concerned with the relationship of their research to society. To be professionally competent, one must be informed and concerned about ethical issues. One needs both a knowledge of ethical standards and the ability to apply them in individual cases.

There are some universal ethical rules for scientists, such as the requirement to do research competently and report the results accurately, and to never harm research participants unnecessarily. In addition to discussing these few absolute rules, we occasionally offer our own opinions on issues; but we mainly describe the important factors each scientist should consider when designing and evaluating a study. We believe that ethical decisions are best made by educated scientists who understand the value implications of their decisions, carefully consider the moral alternatives when making choices, and then accept responsibility for their actions.

I Ethical Treatment of Participants

2 Harm to Participants

If today we concern ourselves exclusively with the technical proficiency of our students and reject all responsibility for their moral sense, or lack of it, then we may someday be compelled to accept responsibility for having trained a generation willing to serve in a future Auschwitz.

Alvin Gouldner (1963)

The most basic guideline for social scientists is that subjects not be harmed by participating in research. Extreme physical or psychic pain, personal humiliation, and loss of interpersonal trust are examples of harm that could occur in behavioral science studies. There are several important reasons why the ethical investigator wishes to avoid harming subjects. First, scientists recognize that individuals have basic rights that are guaranteed both by our legal system and by our moral values. One of these is the right not to be hurt by others. Second, a major goal of science is to benefit humanity; research that harms subjects would be in opposition to this goal and to the values of scientists. And finally, scientists avoid harming subjects because they recognize that such harm may result in a general distrust of science. If scientists themselves do not guarantee their subjects' welfare, strict safeguards will soon be imposed by society. Despite these compelling arguments for the protection of participants, some researchers have not been sufficiently zealous about minimizing harm. In addition, there are areas in which some risk is difficult to avoid.

Case Studies

To familiarize the reader with the issues, several studies, summarized below, document some of the many types of harm that subjects may experience in research in the social sciences. The reader should keep the following questions in mind. What types of harm, if any, were subjects exposed to in each study? What principles can protect research participants from such harm?

17

FEAR, ANGER AND STRESS

In an early study, Albert Ax (1953) induced moderate anger or extreme fear in subjects so he could measure their physiological responses during emotional states. Subjects were falsely told beforehand that the procedures were designed to measure physiological differences between those with and without hypertension. To produce anger, the experimenter insulted the subject, criticizing him for five full minutes. The anger produced was attested to by one male subject, who remarked to a second experimenter, after the first had left the room, "Say, what goes on here? I was just about to punch that character in the nose."

The fear condition was more dramatic and produced even stronger emotional responses. Subjects were led to believe that the electrodes attached to their bodies were for recording physiological responses, but suddenly they were given an unexpected electric shock through a finger electrode. When they reported the shock to the experimenter, he unobtrusively pressed a button that caused sparks to fly from a machine near the subject. The experimenter then excitedly shouted that this was a dangerous high-voltage short circuit and created an atmosphere of confusion as he scurried about the room. This melodrama was sufficient to produce the desired fear. One woman begged for help and pleaded for the wires to be removed. Another woman prayed to God to spare her. One man stated very philosophically, "Well, everybody has to go sometime. I thought this might be my time." This very stressful study lacked many of the safeguards that are widespread today. Informed consent, pilot studies to assess the stress induced, debriefing, and followups were not mentioned in the published report.

Other extremely stressful procedures have also been employed in behavioral studies. In a series of studies on performance during stress, army scientists induced extreme stress in young soldiers by leading them to believe they were about to die (see chapter 1); and Campbell, Sanderson, and Laverty (1964), in a study on conditioning, made alcoholics think they were about to die by paralyzing their muscles with a drug (see chapter 3). Another very stressful study, by Lindzey (1950, p. 301), is best described in his own words:

> The frustration experience involved subjecting the subjects to some ten to twelve hours of food deprivation, inducing them to drink from a pint to a quart of water and preventing urination for approximately three hours, taking a blood

sample with a sterilized spring lancet in such a way as to cause considerable pain, and, finally, forcing them to fail at an assigned task in a group situation.... In addition to all this, he was subjected to approximately two hours of a long and tedious task in a tense atmosphere.

The extremely strong negative emotions produced in these and similar studies present an ethical problem. Subjects were probably in little danger of physical injury, although heart attacks and other physical accidents were possible. But the risk of emotional harm is the reason for questioning whether these studies should have been conducted as they were.

There are many other stressful studies that probably represent little or no risk of permanent harm to subjects. For example, Lazarus (1964) showed subjects films of an Australian tribe in which men perform a crude ritual operation on the genitals of adolescent boys; as the subjects watched the film, physiological measures were recorded. Hess (1965) showed participants pictures of piles of bodies in a concentration camp, then measured their pupillary reflex to the distressing scene. Mathes and Guest (1976) asked subjects if they would carry a sign around the college cafeteria reading "masturbation is fun." Subjects were not actually required to carry the signs. These studies illustrate the question whether experiments are ethical if they produce strong temporary stress or embarrassment but are unlikely to lead to permanent injury.

INDUCING SUBJECTS TO PERFORM REPREHENSIBLE ACTS

Entrapment studies are those in which participants are covertly encouraged to perform illegal or immoral acts. West, Gunn, and Chernicky (1975) tempted people to participate in a burglary; the object of the research was to apply a prediction from psychological attribution theory to a situation like the Watergate break-in. The hypothesis, derived from a theory about the difference between the perceptions of actors and onlookers, was that those participating in the burglary would perceive it as less reprehensible because of their active role. The hypothesis was tested by having a private detective contact criminology students and present an elaborate plan for burglarizing a Florida business. In one condition subjects were told they would be paid $2,000 for their part in the crime. Many of the subjects agreed to take part, but the researchers of course did not carry out the crime.

Besides the extensive deceptions involved, entrapment experiments may harm subjects in several ways. Participants are sometimes enticed to perform apparently immoral or illegal acts, and encouraging such wrongdoing is highly questionable. Moreover, the subjects' feelings of self-worth can be lowered when they learn the true nature of the study. In the "Watergate study," for example, the subjects who had agreed to participate in the burglary undoubtedly realized after the debriefing that they were being compared to those who "went along" with the Watergate break-in, which was being heavily criticized nationwide. Subjects may have felt guilt after agreeing to participate in a similar illicit activity. Some may argue that it is helpful to participants to reveal to them that under certain conditions they could be induced to commit illegal acts: it may help them resist pressure in the future. Articles by West and his associates (1975), by Cook (1975), and by Schlenker and Forsyth (1977) provide further discussion of the ethical issues involved in the "Watergate study."

Entrapment studies were also conducted by Milgram and Shotland (1973) to research the effects of television crime on real-life behavior. In their first study, subjects in a theater were shown one of several versions of "Medical Center," a television drama featuring a "Dr. Joseph Gannon." In one version a character broke into a charity donation box and stole the contents; in the control version, no theft took place. All subjects were later given an opportunity to steal about fifteen dollars from a Project Hope donation container; they were allowed to keep the money if they stole it and were never debriefed after the study. In later versions more prompts to stealing were added, first by frustrating participants before the theft opportunity, then by placing a March of Dimes container, which appeared to have been broken into, near the intact Project Hope container to serve as a model of antisocial action. The television crime did not seem to increase subjects' antisocial behavior over that displayed in the control group. The investigators, having thus found that watching television thefts had virtually no effect on real-life stealing, moved to Detroit and Chicago and switched the crime portrayed to abusive phone-calling. On the "Medical Center" segment aired in Detroit a man was portrayed making abusive phone calls to a charity telethon, whereas in the control program, broadcast in Chicago, the abusive phone-call scene was deleted. During the commercial breaks a public-service message requested viewers to phone in pledges to an authentic charity, Project Hope. Operators were on hand to record pledges and also to write down any abusive calls triggered by the episode of "Medical Center." In order to stimulate calls, ads for Project Hope were placed in the Chicago and Detroit newspapers for several days just before and after the program was aired. The abusive-call experiment was repeated in New York. In none of the situations did the television program seem to stimulate real-life abusive calls.

The Milgram and Shotland studies clearly illustrate the ethical dilemmas researchers face when doing entrapment studies. The questions asked were of great importance because of the many crimes shown on television. The investigations were realistic because actual television programs and antisocial behaviors were employed. As it turned out, the television sequences did not encourage real-life antinormative behavior, which was an important finding. But imagine for a moment that this had not been the case. How many suggestible persons may have been negatively influenced by the program, and how many innocent persons and unrelated charities may have been affected? There is certainly a potential for harm, both to subjects and to society, in entrapment studies of this type, in which wrongdoing is encouraged. The resultant harm may take the form of loss of self-esteem or a change in the morals of the subject during the study. On the other hand, the crimes shown as part of the experiment represented a trivial increase in the number of antisocial acts that are aired on television every year.

A BLOW TO SELF-ESTEEM

A number of investigators have attempted to influence subjects' self-esteem. Their studies further demonstrate the potential for psychological harm, which in the social sciences is usually of greater concern than physical harm. For example, Walster (1965) raised or lowered female participants' feelings of self-worth and measured the effect of this manipulation on romantic liking. Female college students were given a personality test followed by phony feedback that indicated either that they had extremely healthy personalities or that they were constricted, unimaginative, and uncreative. Gerry Davison, a handsome graduate student, pretended to be another subject and struck up a conversation with each woman. As they sat in the waiting room together, Davison acted interested in the female subject and told her some of his own background. He then asked her out for a dinner date and show in San Francisco. It was Walster's intention to measure the effect of the manipulation of self-esteem on participants' romantic attraction to Davison. After the experiment, subjects were told of the hoax and were informed that there would be no date.

The potential harm from Walster's study was psychological, not physical. Consider, for example, the phony personality reports. Those receiving negative ones probably felt bad about it, and we do not know whether the debriefing eradicated these negative effects. Those given the positive reports may have been angry or embarrassed at having been fooled, as well as disappointed that their positive personality reports were not genuine. The dating ruse was potentially even more harmful. Many of the women were probably flattered by Davison's interest and excited about the prospective date, but then they learned that his interest in them was not genuine but was just his job. Dating anxiety and low self-esteem owing to lack of dates are a major problem at college counseling centers. For students facing identity crises and severe self-doubts, the revelation of Davison's actual lack of interest in them may have encouraged self-criticism. There are legitimate differences of opinion about the damage this study caused; but Davison himself has since suggested that the study was not worth the harm involved (Rubin 1970).

Is Harm Ever Justified?

Each reader has probably formed his own opinion about whether there was any harm to subjects in these cases and whether the studies were ethically justifiable. There are differences of opinion in the social sciences about what constitutes harm and whether placing subjects in potentially harmful situations can *ever* be justified. One view was voiced by Norman Denzin (1968, p. 502), who stated, "The goal of any science is not willful harm to subjects, but the advancement of knowledge and explanation. Any method that moves us toward the goal, without unnecessary harm to subjects, is justifiable." This implies that harm to subjects is ethical as long as the injury is necessary to the advancement of scientific knowledge. But the criterion that research need only advance science to be ethical can be used to rationalize any study, including atrocities. Thus it should be rejected as offering too little protection for society and research participants.

A more conservative view on harm to subjects is conveyed by the

will indicate why this guide cannot by itself justify potentially harmful studies.

Costs and benefits are impossible to predict. Nobody can accurately predict the benefits of a particular social science study before it has been conducted. Although it can be foreseen that some research will yield nothing of value (often because of faulty experimental design), nobody can predict whether most studies will make small or outstanding contributions. Breakthrough discoveries are so rare that predicting them has been impossible. The very fact that the study is being conducted suggests that the outcome is unknown.

Costs and benefits are impossible to measure. Even in the unusual case where the outcome of a study can be successfully predicted, a cost/benefit analysis is difficult because there is no scale by which human costs and benefits can be measured. For instance, how costly is it if subjects lose self-esteem in a study? It is equally difficult to measure benefits. What benefits accrue if the study helps refine a theory? The difficulty or impossibility of quantifying benefits is clearly evidenced by the fact that there is vast disagreement among social scientists over the worth of many completed studies. There is simply no way to accurately measure the costs and benefits of a study and to balance the two against each other.

Balancing individual costs against societal benefits. In medicine the cost/benefit ratio was initially applied to the individual, with both the risks and the possible gains accruing to that person. Many now have extended this principle to the situation where individuals incur the costs but society in general will receive the gains. It is one thing to expose an individual to risks when he will probably profit; it is quite another to expose people to risks when they will carry the costs and society in general will reap the rewards (M. B. Smith 1976). Of course individuals may make noble sacrifices for the advancement of science or the general welfare, but the researcher can justify exposing people to serious harm only if the risk is clearly explained to the subject and he voluntarily agrees to undertake it (Baumrind 1976; Kimble 1976).

Conflict of interest in the decision-maker. Another basic drawback of cost/benefit analysis is that it must be completed by the researcher, the very person who believes the research is valuable and who has the most to gain from the study. The scientist's intense involvement with the study means he will often be an unlikely source of an objective cost/benefit analysis. Many scientists are likely to overrate the importance of their own work and, because they are so close to the decision, to underestimate potential dangers.

Substantive rights. The cost/benefit analysis seems to ignore individual rights in favor of practical considerations. But substantive rights, not utilitarian calculations, pervade current philosophical and legal thinking (Wallwork 1975). The idea of substantive rights is simply that all people have certain inaliable personal rights that cannot be withdrawn for pragmatic reasons. In contrast, pragmatic calculations are based on what course of action will maximize benefits to the majority. The rights to life, religious freedom, and free speech are examples of rights that may not be ignored even though society might function more efficiently if they were withdrawn. For example, the famous Miranda decision guaranteed that upon arrest individuals must be clearly informed of their legal rights. Because of this decision some convicted criminals have had to be released from prison. A cost/benefit analysis is inadequate by itself because such a calculation fails to recognize that some human rights are inviolable.

Cost/benefit analyses reconsidered. There appear to be several severe drawbacks that make the cost/benefit guideline inadequate by itself, even though many people use cost/benefit considerations in judging the ethics of a study (Edwards and Greenwald 1977). These considerations should nevertheless be reviewed as a first step in an ethical analysis. This analysis cannot by itself justify an investigation, but it can be sufficient cause to abandon a study. For example, a study that is badly designed is not likely to benefit anyone, even the researcher. Poorly designed research is therefore in a sense unethical, since the projected costs must exceed the projected benefits.

INFORMED CONSENT
The traditional biomedical criterion (United States Department of Health, Education, and Welfare 1971) for implementing risky experiments states that subjects may be subjected to risk of harm

26

only if they *understand* the risks and voluntarily *agree* to be exposed to them. As is reviewed in chapter 3, informed consent is an ethical requirement in situations where participants will be subjected to substantial dangers. However, like the cost/benefit analysis, informed consent is not sufficient by itself to ethically justify a study. Informed consent, for example, cannot be used to excuse a study where the risks are much higher than they need to be, nor can it be used to justify a study that is methodologically unsound. Informed consent can be a helpful safeguard because it allows subjects to protect their own best interests, but like the cost/benefit analysis it is not sufficient by itself. Informed consent and the cost/benefit analysis in combination provide considerable ethical protection. However, there are further safeguards that are important in studies where there is a substantial potential for harm to subjects.

CREATING MINIMAL RISKS

Another principle intended to protect subjects states that the risk in an experiment should be the minimum essential to test the hypothesis. This means that experimental manipulations should not be more extreme than necessary and that the scientist should take all possible precautions to protect participants. Levine (1975*b*) has listed several dimensions along which harm may vary. A study will be more ethical to the extent that it falls toward the safe end on each of these dimensions:

1. Likelihood of occurrence. The less probable the harm, the more justifiable is the study.

2. Severity. Of course minor harm is more justifiable than possible major injuries.

3. Duration after the research. The investigator should determine whether harm, if it does occur, will be short-lived or long-lasting.

4. Reversibility. An important factor is whether the harm can be reversed.

5. Measures for early detection. If incipient damage is recognized early it is less likely that people will be exposed to severe harm.

One guideline not listed by Levine is that if the potential harm is similar to the risks of everyday life it may be justifiable. For example, driving a car on the highways is somewhat hazardous. But because driving is so commonplace in our society, having people drive a car as part of a study would normally be justifiable even though it exposed subjects to some risks. These six dimensions should be helpful guidelines to investigators in assessing and minimizing the dangers of a study.

Where it appears that the risks inherent in an experimental manipulation are substantial, an alternative is to turn to field situations where nature produces effects similar to those the scientist is interested in. By relying on natural events, the researcher can insure that the research itself does not create risks for subjects. For example, he could not justify creating brain lesions in humans to assess the functioning of various cortical areas. But nature has produced such lesions in some people through accident or disease, and the scientist may seek them out as volunteers for a study (e.g., Teuber 1960). Similarly, a scientist could not ethically justify depriving children of vitamins to study the behavioral effects of malnutrition. But he might study a similar phenomenon in natural settings by working in areas of the world where children are chronically malnourished (e.g., Klein, Habicht, and Yarbrough 1973). In summary, the scientist has a strong moral obligation to reduce dangers to the minimum consistent with the research goals. Where it appears that direct research interventions will pose serious risks, an ethical alternative is to seek out analogous situations in the natural environment.

Screening Subjects

One method of minimizing risk to subjects is to screen them for the study, selecting subjects with characteristics that make them more resistant to the dangers involved. For example, an investigator interested in experimentally manipulating self-esteem might consider not utilizing young adolescents or mental patients because the prevalence of shaky self-esteem in these populations might make the manipulations more risky. Similarly, a researcher planning to investigate conditioning to electric shock might not use children as subjects because they are less able to understand shock and may be more prone to strong fears than most adults. Investigators might also screen out particular individuals who seem more likely to be harmed or, where no bias is thus introduced, even select "hearty" individuals who seem likely to withstand the experimental treatment without difficulty. However, care must be taken not to give those who are eliminated from the study the impression that they have failed to pass a "test for normality."

An example of prescreening was employed in a simulated prison study (Zimbardo et al. 1973) in which subjects were to role-play prisoners or guards for two weeks. Zimbardo (1973) gave an extensive battery of tests to volunteers and selected only those who fell within the normal range on all tests. By choosing subjects with "normal" personality profiles, Zimbardo hoped to reduce any destructive effects. Practically, researchers seldom need to be concerned that screening of subjects will destroy the representativeness of their sample, since they usually start with atypical samples anyway.

PREDICTING HARM: PILOTS AND ROLE-PLAYERS

Social scientists are frequently unable to predict whether an experimental procedure will be harmful to subjects. For example, some scientists believe that questions about sexual behavior may be most stressful, whereas others think that threats to self-esteem are of greater risk. We often cannot accurately predict how subjects will respond to various types of unpleasant or risky situations. Therefore, when there are potential hazards in an experiment, the ethical researcher may run pilot subjects and carefully interview them afterward for their reactions and suggestions. In addition, Farr and Seaver (1975) have published helpful information on how unpleasant college students believe certain commonly used procedures would be for them. For very risky experiments the first pilot subjects might role-play the experiment, followed by running true "pilots" if the reactions of the role-players suggested that the procedure was safe. In order to avoid exposing the pilot subjects to great harm, the potential negative stimulus may be introduced gradually.

ASSESSING AND AMELIORATING HARM

The last safeguard necessary in potentially harmful studies calls for the experimenter to institute procedures for assessing harm and to plan curative action where any harm is found. Regardless of other precautions, if there is a potential for measurable harm, a careful assessment should be made immediately after the experiment. In more serious cases, a long-range follow-up also may be advisable.

Many experimenters now routinely interview subjects after the session ("debriefing") to ameliorate any negative reactions they may have. At times, a back-up person such as a clinical psychologist has also been available for referral, should any subject become overly upset. Thus far, reports in the literature have not revealed harm to subjects uncovered in immediate postexperimental interviews. It is also encouraging that the few long-term follow-ups that have been done indicate that no long-lasting harm occurred in these studies. Milgram's (1963) famous study using an obedience-shock paradigm has been criticized for ethical reasons by some scientists because most subjects were placed under substantial stress and were coerced into an activity most people consider morally repugnant—delivering supposedly dangerous shocks to another "subject" (Baumrind 1964). Milgram (1964) and Ring, Wallston, and Corey (1970) interviewed subjects immediately after shock-obedience studies and then again several weeks later. Neither found any long-term harmful effects. In fact they found that most subjects thought participating was a valuable experience.

Zimbardo (1973) also used follow-ups to assess possible harm in his simulated prison study that has drawn ethical criticism. Zimbardo created a prison environment in the basement of the Stanford psychology building in which some subjects role-played guards and others acted as "prisoners" (Zimbardo et al. 1973). It turned out that the prisoners became docile and subservient to the guards and somewhat hostile toward one another. Several prisoners broke down and had to be released. On the other hand, several of the guards became cruel and degraded prisoners with verbal abuse and punishment. The study was scheduled to run two weeks but had to be terminated after only six days because of the potentially damaging changes in the subjects. In Zimbardo's own words, "Volunteer prisoners suffered physical and psychological abuse hour after hour for days, while volunteer guards were exposed to the new self-knowledge that they enjoyed being powerful and had abused this power to make other human beings suffer" (1973, p. 243). Zimbardo had not predicted these extreme effects, and after the study he held an encounter session to allow subjects to express their feelings. To his credit, Zimbardo conducted follow-up interviews with subjects to make sure none suffered long-lasting harm from the "prison" experience, and he found no persistent negative effects (Zimbardo 1973). In fact, most subjects reported learning many new things about themselves during their brief imprisonment.

Another follow-up program was completed by Clark and Word (1974) after their study of altruism. The researchers simulated a situation in which a technician received a high-voltage shock that knocked him unconscious. A measure was taken of how many persons assisted the "victim." Sixty-eight percent of the subjects claimed to be upset at the time of the emergency, but after a thorough debriefing, only one subject out of seventy-eight still claimed to be upset. When subjects were interviewed three to six months later, they reported virtually no negative aftereffects from the stressful experiment. These longer-term follow-ups are informative because they did not uncover long-lasting ill effects from what appear to be very stressful studies. One possible limitation of follow-ups is that they may be unlikely to uncover negative effects because subjects are unwilling to admit they were adversely influenced. However, many subjects stated in the follow-up interviews that their research participation was an important and worthwhile growth experience in which they learned about themselves and human behavior. It is noteworthy that these relatively stressful experiments were often regarded as genuine learning opportunities, whereas subjects rarely consider participation in paper-and-pencil studies, which present no ethical problems, to be an enlightening experience.

The lack of negative effects in the follow-ups that have been conducted does not mean there have not been long-lasting negative effects in some studies where there have been no follow-ups. In a study by Campbell, Sanderson, and Laverty (1964), it was found that in some subjects a conditioned fear could not be extinguished. Thus far, however, there is no documented evidence suggesting permanent damage from participation in social science research.

REVIEW BY OTHERS

Review of a study by colleagues and other concerned individuals is often extremely helpful in minimizing risks to subjects, for other persons may recognize dangers the investigator did not foresee and recommend safeguards that did not occur to him. Indeed, under a recent law enforced by the Department of Health, Education, and Welfare (Weinberger 1974, 1975), much research at institutions within the United States must now be reviewed by a formal institutional review board. This law mandates specific safeguards where participants are deemed "at risk." Institutional review boards and consultation with colleagues can help researchers decide how much risk is likely in a study. Even when formal review is not required, the wise investigator will seek out the opinion of others. Barber et al. (1973) recommend that *all* research be reviewed and

31

that review committees contain some persons from outside the profession.

Guidelines Summarized: Protection of Subjects in Risky Experiments

It is encouraging that so far there is no documented case of a subject's being seriously harmed by participating in social science research. In fact it appears that most social science studies are innocuous and do not represent risks more serious than those that are part of everyday life. Nevertheless, some studies do contain potential dangers to subjects. In choosing whether to conduct these possibly risky investigations, researchers can use several ethical principles to guide their decisions:

1. Do the benefits of the study outweigh the costs? Although this criterion is certainly not sufficient to justify a study ethically, investigators should proceed only if the study appears to meet it. Thus cost/benefit analysis is a necessary but not sufficient ethical guide.

2. Has informed consent been obtained? Participants must be informed of serious risks before the study begins. Where the potential harm is very minor or unlikely to last beyond the brief experimental session, informed consent may sometimes be ethically omitted if it would be detrimental to the research design.

3. Will all possible precautions be taken to keep risk of harm to the lowest possible level, while still allowing the study to be effectively completed? Few studies in the social sciences (at their present level of sophistication) promise to provide major scientific breakthroughs, and therefore risk of substantial harm can rarely be justified in our research. When serious damage to subjects is likely, the investigator may turn to the everyday world where the variables of interest may occur naturally.

4. Will an effort be made to select subjects who are unlikely to be harmed by the experimental treatment? This safeguard has been underused. Some researchers choose subjects according to convenience, without considering whether the group may be highly susceptible to the harm posed by the study. More effort should be made to use careful subject selection as an ethical safeguard.

5. If an investigator is unsure whether the study will have harmful effects, will role-playing and pilot subjects be run before the full study is conducted? The pilot subjects should be carefully inter-

viewed to uncover information about how the study can be made less risky. The investigation should be canceled if the pilot study suggests that the dangers are substantial.

6. Where injury to subjects is possible, will measures be instituted to assess harm after the study? Where the harm may be long-lasting or may appear only after some time, a follow-up interview should be arranged. Of course, where harm is uncovered there is a serious ethical obligation to do one's best to eliminate it.

7. Will consultation from colleagues or an ethical review board be sought for possibly risky studies in order to gain additional insight and a more disinterested perspective?

For further reading on evaluating risk to subjects, the reader is referred to Barber 1975; Baumrind 1976; Farr and Seaver 1975; Katz 1972; Levine 1975*b*; National Academy of Sciences 1975.

3 Informed Consent

The cornerstone of all considerations of the welfare and protection of subjects appears to be what has been called informed consent. This term refers to a person's ability to consent freely to serve in an experiment in which he adequately understands both what is required of him and the "cost" or risk to him.

Wolf Wolfensberger (1967)

While the informed consent of subjects should be obtained whenever feasible ... it should not be considered a ritual that absolves the investigator or the research community of all responsibilities.

Paul Reynolds (1972)

Informed consent is the procedure in which individuals choose whether to participate in an investigation after being informed of facts that would be likely to influence their decision. Informed consent includes several key elements: (*a*) subjects learn that the research is voluntary; (*b*) they are informed about aspects of the research that might influence their decision to participate; and (*c*) they exercise a continuous free choice to participate that lasts throughout the study. The greater the possibility of danger in the study and the greater the potential harm involved, or the greater the rights relinquished, the more thorough must be the procedure of obtaining informed consent.

Informed consent is absolutely essential whenever subjects are exposed to substantial risks or are asked to forfeit personal rights. The United States Department of Health, Education, and Welfare guidelines governing research supported under its grants indicate that a signed consent form should be completed if subjects are placed "at risk" (Weinberger 1974, 1975). A possibility of danger does not preclude important research but does necessitate the use of informed volunteers. When participants are to be exposed to pain or stress, to an invasion of their privacy, to emotional, physical, or sociological injury, or when they are asked to temporarily surrender their autonomy (e.g., in drug or hypnosis research), then informed consent must be fully guaranteed. No doubt should be left in subjects' minds that the research continues to be voluntary, and they should receive a thorough explanation beforehand of the rights and dangers involved. Informed consent procedures should also be instituted when subjects are to expend a sizable amount of time or effort in the study.

In research that poses no risks to participants, the absolute necessity of informed consent is more questionable. Some recommend that informed consent be mandatory in all research (e.g., Veatch 1976), but the requirement that informed consent be obtained even in studies that are entirely harmless is probably unrealistic and unnecessary. As we shall see in later sections, an absolute requirement for informed consent in all research is not needed to protect subjects and would make some types of studies impossible (e.g., unobtrusive field observations).

Why Informed Consent?

The principle of informed consent is based upon both cultural values and legal considerations. Informed consent partly rests on the high priority assigned to freedom in Western societies. Milton Rokeach (1973) found that Americans rank individual freedom, along with peace and family security, as one of their most esteemed values. It is clear that informed consent serves this value. For a behavioristic view of freedom, see Vargas (1977).

In Western law there is a heavy emphasis on freedom of choice and lack of coercion. Both the constitution and the common law within most democracies guarantee people a large number of freedoms and rights, such as freedom of speech and privacy. Those who trespass upon these rights are often subject to criminal prosecution or civil suit unless the person affected has voluntarily relinquished them. Thus people have the legal right to give or withhold their informed consent in situations where their rights might be jeopardized.

Reflecting the high valuation of freedom in Western culture, almost every code of professional ethics affords informed consent a central place. For example, the first principle in the Nuremberg Code (e.g., 1964) states, "The voluntary consent of the human

35

subject is absolutely essential." And in the social sciences, a key principle of the American Psychological Association states: "The individual's human right of free choice requires that the decision to participate be made in the light of adequate and accurate information (1973, p. 27)."

In addition to the philosophical and legal considerations concerning freedom of choice, there is a simple "commonsense" justification for informed consent: the idea that people's rights and welfare are usually best protected when they themselves make the decisions about things that directly affect them. Because people will usually protect their own interests, allowing them freedom of choice about participating builds into the research a safeguard against extremely hazardous procedures. If research subjects must be convinced that participation is both safe and worthwhile, then experimenters will be careful to implement safeguards and minimize hazardous procedures. Informed consent thus serves to reduce potentially harmful research and guarantees that subjects will be exposed to danger only if they voluntarily agree to it.

Another reason for the concern with informed consent is that it can improve the quality of the subject-experimenter relationship. Where subjects are given no freedom of choice about whether to participate in research, they may truly become "guinea pigs," like experimental animals or objects to be *used* by the researcher. On the other hand, when subjects voluntarily agree to participate after being fully informed about the study, their participation can be a positive expression of man's values. The desire to make a contribution to human welfare and the motivation of intellectual curiosity, combined with freedom of choice, can make the research experience beneficial for both the participant and society. For this reason most subjects will consider informed consent valuable even when no risks are involved.

An example of a study without informed consent will reveal how subjects may be treated with little respect when they are deprived of freedom of choice. Campbell, Sanderson, and Laverty (1964, p. 629) wanted to assess the effects of traumatic conditioning. The "volunteers" were hospitalized alcoholics who were told that the procedure was "connected with a possible therapy for alcoholism." In fact, the study was not designed to treat alcoholism and the researchers had no reason to believe that the conditioning procedure would cure the volunteers. The method used to create traumatic conditioning paired a neutral tone with intense fear in a classical conditioning paradigm. This procedure makes the neutral tone able to cause fear by itself after being paired with the cause of fear. The "volunteers" heard a tone and then were injected with Scoline, which produced motor paralysis, including paralysis of the diaphragm, so that the patients could not move or breathe. The inability to breathe for an average of almost two minutes produced real terror. Even though no permanent physical harm resulted, all the alcoholics in the experimental group thought at the time that they were dying. The procedure produced a long-lasting, conditioned fear reaction to the sound. Whenever the tone was presented the subjects reacted with fear. The fright reactions could not be extinguished in some patients even after a large number of trials.

This study illustrates a breach of informed consent, not only because subjects were apparently unaware of the pain, risks, or extreme stress entailed in the study when they volunteered, but also because the experimenters apparently lied to the subjects about their interest in a possible cure for alcoholism in order to obtain "volunteers." To correct these problems of informed consent, the experimenters could have asked for volunteers for a study on conditioning. They should have outlined the fear-producing effects of the drug and the possible risks. If the participants still agreed, once they understood about the study, the researchers could have proceeded. In fact, several physicians did volunteer for the study, knowing about the drug's terrifying effects. They too acquired the conditioned fear, which reveals that ignorance about the drug was not necessary for the success of the experiment. Thus the alcoholic subjects were deceived for no purpose other than to gain their uninformed cooperation.

In summary, there are important reasons why informed consent is desirable in most research and absolutely essential whenever substantial hazards are involved:

1. Informed consent increases subjects' freedom to choose whether to participate and guarantees that exposure to risks is undertaken voluntarily.

2. Informed consent protects subjects by screening out those who feel they might be harmed by the study and by stimulating investigators to create safeguards.

3. Informed consent tends to increase societal trust and respect for science by creating a joint enterprise between experimenter and

subject. Furthermore, most subjects will value informed consent even when no risks are involved in the research.

4. Informed consent reduces the investigators' legal liability because subjects have voluntarily agreed to undergo the procedure.

Obtaining Consent

LABORATORY STUDIES

In laboratory studies it is easy to request informed consent at the beginning. When participants are not exposed to risk, a brief oral description is sufficient to inform them about the study. Out of respect for participants' freedom they should be told that they may withdraw at any time. If the subject does not desire to withdraw, his agreeing to proceed is a sufficient indication of consent. Where subjects forfeit substantial risks or personal rights, the oral explanation should be supplemented by a written consent form to be signed by each subject, containing a brief description of the study and the risks involved and a statement that the individual may withdraw at any time. A sentence to the effect that the subject has read the information and voluntarily agrees to participate should appear at the end, immediately before the subject's signature.

SURVEYS

For surveys or interviews in field situations, explanations are simple but important. Survey interviewers often precede their visit with a postcard and usually show identification to assure people that they are not salesmen in disguise. Interviewers next should explain what the survey is about, how much time it may take, and who they represent. It is usually valuable to tell people how they were selected and to assure them that their responses will be completely anonymous. A willingness to answer the questions is usually adequate assurance of informed consent in survey or questionnaire studies, but people should be informed that they may choose not to answer sensitive questions (e.g., about income) if these are included in the survey. Whereas most surveys are open, occasionally disguised interviews are used in which people do not know that their responses are being recorded (e.g., Perrine and Wessman 1954). These cases may violate subjects' right to informed consent.

FIELD OBSERVATIONS

The clearest cases where informed consent is irrelevant for behavioral science studies are those that involve the observation of public behavior or public records. Picture, for example, an eager sociologist poring over courtroom records for a study on the influence of ethnic status on trial outcomes. It would be unnecessary for him to obtain consent from the accused to use information that was already recorded in public documents. Or imagine the field setting employed by Bryan and Test (1967) on donation behavior during the Christmas season, in which both black and white Santas rang bells next to Salvation Army buckets. Every busy shopper who hurried by, whether or not he donated, was actually a subject. Picture the irritation of the harried shoppers if, instead of telling them nothing, the experimenters had requested informed consent from each of them. The Santas and donation buckets represented no risk to subjects, did not infringe on any rights, and were wholly within the experiences of everyday life. And the donations actually went to the charity.

When field studies do not significantly affect subjects' lives, informed consent becomes irksome and time-consuming for all parties and may be both ethically and methodologically undesirable. The investigator could scarcely justify waylaying people in public places to describe a study that had little or no effect on their lives and of which they were totally unaware. Gaining informed consent would require a substantial amount of the experimenters' and subjects' time, probably more time than the observation took. Furthermore, the move to the field is often motivated by a desire to observe spontaneous and natural behavior. If informed consent were obtained first, spontaneity would probably be destroyed. Where public behavior is observed and subjects are exposed to no dangers or substantial costs, it is normally sufficient to clear the study with reviewers (for an example, see also *Anthropology Newsletter* 1977).

When investigators perform experimental manipulations on subjects in the field or alter the natural environment to influence people's behavior, questions of informed consent begin to arise, but even in these circumstances there are many studies in which consent is unnecessary. For example, Merritt and Fowler (1948) developed the "lost-letter technique" to test the honesty of people in various cities and locations. They dropped two kinds of letters: one type contained only a letter and another contained a slug the size of a fifty-cent piece. Eighty-five percent of the plain letters were returned, but only 41 percent of the letters with slugs were returned

39

unopened. Field methods such as the lost-letter technique that are not risky, do not invade subjects' rights, and require only small amounts of time usually do not require informed consent. Where the behavior of subjects is affected in minor ways (for example, choosing to return a lost letter to a mailbox or answering a wrong number), and when the simulated event is commonplace in everyday life, the researcher may justify omitting individual informed consent. However, it is best if the researcher checks the study with a sample of persons similar to those who will be studied to insure that they do not object to the procedure.

Incursion into people's right to privacy illustrates the first type of field study in which informed consent is desirable. Very private behavior should be studied only when subjects agree to participate after learning about the nature of the study. For example, Masters and Johnson (1966) observed people during the most intimate sexual encounters, but only with the participants' full knowledge and consent. Similarly, the private behaviors of families have been recorded with their knowledge and permission (e.g., Baldwin 1949). Since subjects voluntarily relinquished their right to privacy within their home, the potential ethical problems about privacy were thus avoided. Similarly, if people's pictures are taken as part of a study, their permission to use the pictures should usually be obtained afterward. If a subject objects, his picture should be destroyed. If the picture is of a large group, consent is often not necessary but may be desirable in the case of deviant groups with whom subjects might not want to publicly identify. If the individual is recognizable in a picture, his permission should be obtained before the photograph is published.

When people in field studies are subjected to possible harm without consent, serious ethical problems arise. Some emergency bystander studies, in particular, can be questioned from this perspective. For example, researchers have faked collapses on subways and let blood trickle from their mouths (Piliavin and Piliavin 1972). Such events are not common in the lives of most subjects and may expose them to stress and risk of harm. For example, it is conceivable that upon seeing a person who is apparently bleeding internally, an onlooker could have a heart attack. And it is possible that a mentally unstable person could be unnerved by these faked scenes. Even the healthiest of subjects could be injured in a rush to help. If no one helps, the self-esteem of the individuals may be lowered when they reflect on what they should have done. Obviously, in a study like this informed consent is impossible to obtain and the investigator must carefully consider the risks involved. If risks are small and precautions are taken to insure the rights of subjects (such as clearance with group representatives), the experimenter may often proceed after having the study reviewed by others.

Another ethical problem faced in field research arises when subjects incur substantial costs, including the expenditure of a sizable amount of time. For instance, when researchers pose as customers in an empty store, the costs to the employees are not great, since salesclerks deal every day with many customers who are only window shopping. But if investigators were to take up large amounts of time during busy sales hours, they would be costing the store and the clerk both time and money. In such cases the investigators might obtain consent in advance from the representatives of the group or, even better, from the individuals themselves (see procedures on p. 45). They could also reimburse the subject for his time or costs. Another alternative in some studies is to obtain consent post hoc and destroy the data if the subject desires. Through expedients like these, the investigator, not the unknowing participant, is made to pay the costs incurred *in the study*, and it is the researcher who must decide if the possible knowledge is worth the costs.

In summary, the types of field research where informed consent is needed are those where substantive rights of subjects are jeopardized, where subjects are placed at risk, or where subjects incur a substantial cost in time or money. If any of these problems are present in a proposed field study, it may be unethical to conduct it. The investigator must consider the individual case and assess whether informed consent might be possible before deciding whether to proceed. Other examples of field research and some discussion of the ethical issues involved can be found in Webb et al. (1966), Bickman and Henchy (1972), Swingle (1973), and Wilson and Donnerstein (1976). For issues related to consent in cross-cultural and participant observation studies also see chapters 6 and 7; for issues regarding consent in experimental change studies, see chapter 8.

What Is Sufficient Information?

A question that often troubles investigators is, How much information must one give subjects beforehand? There are differences of opinion about what knowledge participants need in order to exercise truly *informed* consent. Levine (1975a) listed eleven types of information that should be included for fully informed consent: (1) describing the overall purpose of the research; (2) telling the participant his role in the study; (3) stating why the subject has been chosen; (4) explaining the procedures, including the time required, the setting, and those with whom the participant will interact; (5) clearly stating the risks and discomforts; (6) describing the benefits of the research to the participant, including inducements; (7) where applicable, disclosing alternative procedures—a requirement more clearly applicable to medical research; (8) offering to answer any questions; (9) suggesting that the subject may want to discuss participation with others—obviously most relevant where substantial dangers are involved; (10) stating that the participant may withdraw at any time and that exercising this right will have no negative consequences; and (11) where applicable, stating that the initial information is incomplete and that further information will be given after the experiment.

To this long list of required information, the American Psychological Association (1973) adds further requirements such as information about the sponsorship of the research, the likely gains in scientific knowledge, and how the data will be used. However, many of these elements are unnecessary or trivial in most social science research (e.g., Levine 1975a). Indeed, some of them might be undesirable in a specific study (e.g., informing children that they have been chosen because they manifest behavior problems). Which types of information are important will depend upon the particular study, and the experimenter must exercise good judgment in selecting what information is most relevant.

The investigator must be careful not to cover up facts that might really concern many participants. On the other hand, giving very detailed information is often unnecessary because to protect his rights a subject needs to know what will directly happen to him and if any risks are involved. Participants should also be informed of any details that would be likely to influence their willingness to participate. For instance, if the sponsor of the study is an institution the respondent might not approve of, (e.g., the Ku Klux Klan) or if the results are to be used to draw conclusions about the respondent's ethnic group, then he should be informed of these details.

A study by Resnick and Schwartz (1973) illustrates a situation in which giving complete information was found to be undesirable. These authors resolved to tell subjects everything about a verbal conditioning study before it began, giving them information that usually is not made explicit. Resnick and Schwartz telephoned participants before the study and gave a lengthy explanation. Subjects were told that the experimenter would reinforce their verbal use of plural pronouns with the response "good" and were informed of many other details about the research. Many subjects never showed up for the experiment. Contrary to the results in most verbal conditioning studies, those who did come were not positively conditioned—they actually went out of their way to avoid using plural pronouns. Subjects rated the experiment as uninteresting, and they may have found it ridiculous. The study showed that telling subjects "everything" can have a destructive effect on research outcomes and can also make participation less interesting to subjects.

One criterion in deciding what information should be given to individuals before a study is what a "reasonable or prudent person" would want to know. This legal term captures the essence of a commonsense approach to the problem. There must be a full and frank disclosure of all aspects that a person cautious for his own welfare would need to know before making a decision. Subjects have an *unconditional right* to know of any potential danger or of any rights to be lost during the study. Dangers should not be downplayed, since subjects often have tremendous faith that nothing bad can happen to them in a study (Orne and Evans 1965). For example, in a study on psychological stress, the experimenter should tell the participants that most subjects in the study will feel stress and that it is conceivable that such stress could have harmful effects on some small, unknown percentage. The experimenter will normally also advise subjects of the safeguards that have been instituted on their behalf. Thus, if the participant chooses to go on, he does so voluntarily, with a clear knowledge of the dangers involved.

In summary, the crucial elements of information that should be included in most informed consent procedures are:

1. A statement that participation is voluntary and that subjects may withdraw at any time.

2. A clear statement about risks or personal rights that are

jeopardized by the research and the safeguards undertaken. This is absolutely essential, and in research that exposes subjects to danger, the informed consent should be documented by a signed form.

3. A description of the procedures of the study—that is, what will happen to subjects.

Other elements, such as those listed by Levine (1975*a*) and by the APA guidelines (1973), should be included as required by the specifics of the particular case. Federal regulations governing Health, Education, and Welfare-supported research in the United States mandate several types of information that must be given subjects who are deemed "at risk" (Weinberger 1974, 1975). Levine (1975*a*) made it clear that each study requires that different types of information be conveyed to subjects. In reference to the many potential elements of information he had listed, Levine wrote: "In fact, in most cases most of these factors and devices will be found inappropriate or unnecessary. Each negotiation for informed consent must be adjusted to meet the requirements of the specific proposed activity and, more particularly, be sufficiently flexible to meet the needs of the individual prospective subject (p. 72)."

Deception Studies

Deception studies are vexing when one is considering informed consent because they frequently rely on what might be called "misinformed consent." However, active deceptions are usually ethically quite different from omissions. For example, omitting the purpose of the study from the informed consent procedure is often justifiable. Telling subjects the hypothesis beforehand would invalidate the vast majority of social scientific research. Consider, for example, a study on whether subjects judge essays supposedly written by a woman inferior to those bearing a male name. Some subjects might be given an essay with a male named as author, and others given the same essay said to be by a female author (Goldberg 1968). If subjects were told the entire purpose of the study, they could easily be influenced to judge the essays differently than they would if they approached them naively. It is normally sufficient to tell subjects that the complete purpose of the study cannot be shared with them until afterward.

Omitting information or forewarning subjects that they cannot be told everything beforehand assumes that none of the omitted information concerns serious risks involved in the research. Of course participants retain the right to withdraw if they do not want to participate without full knowledge. Thus omitting information (e.g., about the purpose of the study) and forewarning subjects about this omission is usually ethically preferable to deception (Baumrind 1976). Veatch (1976) has recommended pretesting a pilot sample to assure that they do not consider the specific omission a violation of their freedom of choice.

Where active deceptions concern details of the study that do *not* affect subjects' welfare, rights, or freedom to withdraw, the deception may not jeopardize informed consent procedures. (However the deception may be unethical on other grounds—see chapter 4.) Forewarning subjects that they cannot be given completely accurate information until the end of the study is not enough to justify studies in which a deception withholds information about a substantial risk to subjects. For a further discussion of consent and deception see chapters 5 and 7.

Judging When Consent Is Necessary

Where informed consent will not harm a study methodologically or ethically, it is usually desirable because it helps insure the freedom and cooperation of participants. However, some studies *require* informed consent, whereas others do not. There are areas between the two extremes where it will be unclear whether informed consent is mandatory. The experimenter may be unsure whether the procedures will really be risky or whether subjects will experience them as very stressful. Informed consent procedures may be methodologically undesirable, yet it may be uncertain whether they are necessary. In such cases it may be desirable to determine before the study whether formal informed consent procedures are necessary. One way to do this is to obtain ethical opinions from reviewers, including members of the subject group. For instance, in the example where researchers took up salespersons' time, clerks in the local community should be asked for their opinion. If the clerks find the study innocuous and there is no risk to subjects, the researchers might proceed without individual consent. If it appears that some of the clerks polled object to the practice because they feel "used" or because it might damage their income, the researchers should not proceed with the study unless individual informed consent is obtained. Obtaining informed consent directly from the salespersons who are to be subjects can be done if they are approached earlier by someone other than the person who will pose as a buyer in the study. By the time the research occurs the clerk should be behaving naturally. However, this approach could cause the sales-person to be anxious or suspicious around real customers.

Another way to gain information on whether informed consent is mandatory in a particular study is what has been called "anticipated consent" (Berscheid et al. 1973), in which persons drawn from the subject population role-play the study and then make various judgments about how they would feel if they were actual partici-pants. The subjects either read a detailed description of the study or act it out and then judge whether they would be willing to participate. In the study by Berscheid et al., college students were presented with several descriptions of published studies, some of which were stressful and some not. When asked to participate, virtually none of the subjects objected in the nonstressful studies, but the "refusal" rate for stressful experiments was substantial (about one-third of the subjects).

When post-hoc data from stressful experiments (e.g., Ring, Wallston, and Corey 1970) are compared with the data collected by Berscheid et al., it appears that "anticipated consent" may give an inflated estimate of the number who would be disturbed by the actual study. When debriefing was included in the stressful experi-ments, the post-hoc refusal rate dropped to 11 percent of the subjects. Anticipated consent was proposed by Berscheid et al. as an alternative to informed consent where experimenters could not tell subjects all about the procedures and purpose of the study before it began. However, since one individual cannot normally give consent for others, it seems that anticipated consent can more usefully be thought of as a technique to tell whether informed consent is really necessary. When anticipated consent reveals that a substantial number of subjects are likely to object to the study, securing consent from the actual subjects becomes necessary.

Special Populations

Some groups such as small children and the mentally retarded are unable to give voluntary consent entirely on their own. In such cases consent must be obtained from those legally responsible for the subjects. In the case of children and severely retarded persons living at home, a parent should give permission for participation, prefer-ably in writing, after reading or hearing details of the study. It is best to obtain the subject's permission also, and the researcher should normally not proceed if the participant himself does not want to take part, unless the research promises to substantially and directly benefit the subject (e.g., a new type of therapy). Where subjects are in schools or other institutions (e.g., prisons, homes for the retarded) it is usually necessary to obtain permission for the study from the organization. To use schoolchildren as subjects, it is necessary to obtain permission from the parents and the school. The ideal is to gain permission from a person or agency concerned for the subject's welfare and with no investment in the research.

In the past, the consent of the children themselves has often been neglected as long as parental permission was received. It is ethically superior in most cases to obtain the child's consent as well, especially since he is also concerned with his own welfare and may have a good idea of what would be upsetting. Of course, for infants and preverbal toddlers, the consent of the parent must suffice. But some degree of consent can usually be obtained even from pre- schoolers if what is to be done is described simply, at the child's level of understanding. As children grow older they can understand more about the research and should be given accurate information phrased for their intellectual level.

Parental permission is normally necessary for a minor's participa- tion in research, but it is not always sufficient, and informed consent procedures should be used that allow the child the maximum freedom of choice possible at his age level. For example, Koocher (1974) conducted a study of children's ideas about death. Before the study, the school principal was concerned that such a topic might unduly upset the children. When permission for the study was finally granted, Koocher talked to each child. All the children agreed to participate and were curious about the topic, and none found participation the least upsetting. Koocher followed up on the children after the study and found no negative aftereffects. As in the Koocher study, children will usually be willing to participate if the research does not seem threatening and is described by a friendly researcher. However, a child's reluctance to participate should be respected, even if his fears seem groundless to the experimenter.

It should be noted that in nontherapeutic research that will subject children to substantial risk, it may be impossible to obtain informed consent. Although a parent may volunteer himself for a very risky study, he does not have the right to volunteer his child unless there is a direct and substantial benefit for the child (e.g., in therapeutic research). In other words, one cannot expose children to procedures entailing a high degree of risk except in therapeutic research, because it is impossible to gain completely satisfactory informed consent from either the child or the parent. Further discussion of informed consent with children is available in Annas, Glantz, and Katz 1977; Ethical Standards 1973; Ferguson 1976; Gray 1971; Keith-Spiegel 1976; National Commission 1976, 1978; *Rights of Children as Research Subjects* 1976; Smith 1967. See chapter 8 for additional issues related to consent in experimental change studies, for example studies of new educational methods.

The Problem of Incentives

One special group of subjects, prisoners in penal institutions, illustrates another aspect of informed consent: the issue of which incentives may be strong enough to eliminate freedom in choosing to participate. Early parole cannot be given to prisoners just because they have participated in a research project, for this would violate the ethical responsibility of the correctional staff to protect the rights of society. Prisoners may volunteer for dangerous experiments because they believe it will help their chances of parole, and it is possible that this incentive might be so strong that it would represent coercion to participate. To eliminate the possibility that research participation is not voluntary for prisoners, a number of safeguards may be taken. Participation should not be a precondition for parole, and there should be other ways prisoners can demonstrate "good conduct" (e.g., through work in the prison).

There is a serious ethical problem if refusing to volunteer for research may harm a prisoner's chance for parole, since such a requirement would definitely constitute pressure to "volunteer." Because of the coercive threat implied by parole considerations, it is probably preferable to pay prisoners wages for participating and not enter it in their records for parole at all. Monetary wages that are fair for what is entailed in participation will probably be much less likely to infringe upon prisoners' freedom than would the promise of parole. Experimental participation in itself should never be either necessary or sufficient for parole. But this does not mean that research with prisoners should be prohibited. Indeed, it may be undesirable not to allow the prisoners to give this service to society. After all, experimental participation is one of the few ways prisoners can serve mankind. Not allowing prisoners to participate in research would restrict their civil rights and freedom further by removing from them another right possessed by others in society. Research on the motives of prisoners for participating in scientific studies show that they participate for money, to alleviate the tedium of prison life, and for a variety of other legitimate reasons. They frequently expressed altruistic motives for participation (e.g., to help others), as often as did college students who were queried in the same study (Novak, Seckman, and Stewart 1977). Only 1 percent of the prisoners said they would refuse to be in an important medical experiment. The volunteer rate among those who had been in prison for some time, many of whom had been in previous studies, was very high, indicating that prisoners had found previous experimental participation rewarding. For a further discussion of research involving prisoners, see Annas, Glantz, and Katz 1977; National Commission 1977; National Academy of Sciences 1975; Opton 1974; Schwitzgebel 1968; Veno and Peeke 1974.

Subjects choose to participate in research for many reasons: intellectual curiosity, course requirements, monetary payment, bonus points in a college class, and the advancement of science (Novak, Seckman & Stewart 1977). Since everyone is motivated by some incentives, it would be nonsense to disallow inducements for research participation. Nevertheless, most persons who are concerned with research ethics agree that incentives to participate in research ought not be too strong. In actual practice, however, it is quite difficult to tell what incentives are "too strong." For the compulsive college grade-getter, two bonus points might be an overwhelming temptation. For the very poor, a few dollars might be too strong. It is often difficult to assess ahead of time which motivations may be "too strong." One way of determining whether an incentive is too strong is to see what percentage of people respond to it. If virtually all potential subjects agree to participate even when risks are involved, the incentive may be too strong. However, in practice behavioral scientists rarely have the power or money to offer overwhelming incentives. Another approach is to empirically study the strength of various incentives, since we are currently ignorant about what incentives might be too strong (Reiss 1976).

A very important issue is not the incentives themselves, but the researchers' obligations to act responsibly. For instance, the college professor cannot indiscriminately offer bonus points for experimental participation, because this might conflict with his obligation to be a responsible instructor. The grade distribution might be made up without considering bonus points so that nonvolunteers will not be at a disadvantage, and the instructor should offer bonus points only to the extent that research participation will be a learning experience for students. Similarly, there is nothing inherently wrong with paying poor people for research participation. Participating for money is just as legitimate as the motivation "to help science" that only the better-off can afford. Although it would be wrong to offer poor people such a large amount of money that they virtually cannot refuse, such incentives are usually impractical. The ethical investigator will not use overwhelmingly strong incentives, but neither will he deny the opportunity of research participation to anyone with ethically legitimate motives as long as the motives are not overpowering and do not require the investigator to use incentives that would be wrong for considerations outside the research (e.g., unfair grading).

Limitations of Informed Consent

Despite the desirability of gaining informed consent in most research, there are problems with its *automatic* use in all social science studies (e.g., American Sociological Association 1977; Helmstadter 1970; Reiss 1976; Reynolds 1972; M. B. Smith 1976). The concept of informed consent was originally developed for biomedical research, and laws requiring consent were formulated mainly with this in mind. But even in medicine there are severe drawbacks to overreliance on informed consent. For example, most subjects cannot be truly informed about the degree of risk in a biomedical study because few people have enough knowledge of medicine to really understand the probable risks. This same difficulty in informing subjects of probable risk occurs in the social sciences, and there are other parallel shortcomings. For example, an ill-effect of overreliance on informed consent is the possibility that unscrupulous investigators will shift responsibility from themselves to the subjects. That is, the researcher might use informed consent to justify studies without maximum safeguards, reasoning that the subject can choose not to participate. This *caveat emptor* approach has been rejected as irresponsible in other areas of life such as business, and it cannot be allowed in research. Thus, informed consent is a necessary *but not sufficient* ethical precaution in risky studies. The investigation must still have sufficient merit to justify the risks, and the scientist must take all possible precautions to minimize the dangers. The important point is that informed consent cannot be used to excuse other unethical practices; it is only one of several ethical precautions that researchers must consider.

On the other hand, informed consent should not be made an absolute requirement for all behavioral research just because it is accepted practice in medical research (*American Sociological Association* 1977). As the review by Reynolds (1972) makes clear, the risks involved in social science research, in contrast to medicine, are usually trivial. In the overwhelming majority of behavioral studies subjects are not exposed to any greater dangers than they undergo in the course of everyday life. As Reynolds points out, the usual results in behavioral research are "no effect" or "temporary and mild discomfort" to the subjects that terminates at the end of the study. "Unusual levels of temporary discomfort," such as high anxiety or fear, are rare in social science investigations, and "risk of permanent damage" is virtually nonexistent. Indeed, there has never been a documented case of permanent harm as a direct result of behavioral science research. Because most social science research is nonharmful, it makes little sense to automatically impose strict guidelines of informed consent on every behavioral study regardless of its potential risk. The small amount of time spent in some behavioral studies and the improbability of risk stand in marked contrast to the sometimes life-and-death nature of medical experimentation. Another difference between biomedical and social research is that many social science studies deal mainly with groups, not individuals, and therefore a model of individual informed consent may be difficult to implement or may be inappropriate in protecting participants' rights.

Another reason fully informed consent need not be applied universally is that it may actually have negative effects in some cases. One example of this is the situation where people would be informed about why they have been selected as subjects, and the criterion is in fact socially undesirable in the wider society, for example homosexual tendencies (Baumrind 1971). A researcher could hardly justify informing subjects that they were selected for a study because they are latent homosexuals, or neurotic, or retarded. Although such a procedure would be required by the concept of full information disclosure, it would be unethical in many cases. Researchers should simply omit mention of such negative selection factors. One more problem with a simple model of informed consent applied to social research is that it is frequently unclear who should give consent for a particular piece of research information (Reiss 1976). In sum, informed consent has limitations as an ethical safeguard. In some social science research it is not a sufficient safeguard by itself, while in other studies it is not necessary and may even be undesirable or inappropriate.

Conclusion

Margaret Mead (1969, p. 361) has presented the anthropological model as the ideal of informed consent in the social sciences. She states, "Anthropological research does not have subjects. We work with informants in an atmosphere of trust and mutual respect." Mead emphasizes that anthropologists cannot do their work without the active assistance of their participants. The anthropologist must work and live among those he studies and depends upon their goodwill to gain the desired information. The informants are rewarded through material benefits or through satisfaction of their natural curiosity. In the "anthropological partnership," members of the culture studied help define the important problems and are often enthusiastic about the opportunity to explain their culture to others. Seen in this light, research is a joint venture in which the cooperation exceeds that required by "informed consent." Such a relationship is an ideal that ethically far surpasses the more legalistic conception of informed consent. This approach stressing cooperation and the self-determination of subjects is usually desirable whether or not the study involves risks. The anthropologists' attitude of cooperation can be usefully extended to improve and broaden the approach to informed consent in much social science research.

Informed consent, although usually desirable as indicated above, is not absolutely necessary in studies where no danger is involved. But the more serious the risk to subjects, the greater becomes the obligation to obtain informed consent. Where the risks are foreseeable and considerable, yet informed consent would ruin the study, the research simply becomes unfeasible because subjects cannot be exposed to substantial risk without their consent. If considerable harm is possible, it is not enough that the researcher tries to protect subjects. If there is a danger to subjects that far surpasses what most people experience in everyday life, then in addition to the other safeguards, informed consent is essential.

For further detailed discussions of informed consent the reader is referred to Annas, Glantz, and Katz 1977; Baumrind 1976; Ferguson 1976; Levine 1975*a*; Reiss 1976; and Veatch 1976.

4 Privacy

Inside a sorority house on the campus of a large university, forty women make their way into the chapter room. The atmosphere is steeped in tradition. No one outside their sisterhood is allowed to enter the room, let alone participate in the discussions, which range from upcoming socials to sexual problems. Debbie is particularly attentive during these meetings, watching and listening closely. Unknown to the others, Debbie is recording data on group dynamics from this meeting that will be handed over to a psychology professor the next morning. The interactions will be carefully analyzed and organized for publication. When the study is published, each sorority sister will be given a pseudonym to protect her anonymity. Nevertheless, the sorority members will be able to identify themselves and each other. They will be furious.

The sorority study is fictitious; yet similar designs are employed by researchers in the social sciences. A right has been invaded that demands attention: the right to privacy, "the claim of individuals, groups or institutions to determine for themselves when, how, and to what extent information about them is communicated to others" (Westin 1967, p. 7). Ruebhausen and Brim (1966, p. 426) have defined privacy in a similar way as "the freedom of the individual to pick and choose for himself the time and circumstances under which, and most importantly, the extent to which, his attitudes, beliefs, behavior and opinions are to be shared with or withheld from others." Invasions of privacy are of great concern in an age when there are incursions from many sides into our private lives and when information is easy to obtain. Social scientists may violate privacy by probing a person's private behavior or his thoughts and feelings. Privacy is a value that must be carefully considered when planning research, because the goals of the social scientist will often conflict with this right.

Since the value placed on privacy has varied greatly during

Western history and it has not always been considered an important "right," the reader may wonder why researchers should be so careful to protect it. The very fact that people value the privacy of their thoughts as well as their behaviors is reason enough for social scientists to be careful about potential invasions. We respect other people's values for ethical reasons and also for the pragmatic reason that society will censure science if it tramples cultural values.

In addition to being a cultural value in itself, privacy serves to protect other deeply held values in our society. The freedom of the individual and recognition of his uniqueness and dignity are enhanced by a respect for privacy (Shils 1959). Several arguments can be made to substantiate this claim. If people were forced to reveal all personal information, including facts about their inner selves, they would be highly vulnerable to others who might use this information to influence or control them. The privacy of ideas and feelings protects individual uniqueness by allowing people to believe different things. If people had to reveal all their thoughts, there would be great pressure toward conformity, with a lessening of intellectual and cultural diversity. We rarely share our deepest feelings and thoughts with strangers or mere acquaintances, but we can reveal ourselves to those with whom we are very close. This selective privacy enhances the value of the intimacy we share with a few persons. Privacy serves to protect freedom of thought and behavior, individual diversity, and the boundaries of intimate relationships. The next question that arises is, What information is private?

The Dimensions of Privacy
SENSITIVITY OF INFORMATION
Privacy varies along several dimensions. The first of these is the sensitivity of the information: how personal or potentially threatening it is. For example, imagine that you have received a questionnaire in the mail, asking you to reply to questions about your sexual behavior. Such information is at the sensitive end of the continuum in our culture. An inquiry asking your opinion on proposed housing legislation would prove much less threatening, and a questionnaire asking your opinion of various brands of lawn mowers would probably not be sensitive at all. As the ethical guidelines of the American Psychological Association (1973, p. 87) state: "Religious preferences, sexual practices, income, racial prejudices, and other

55

personal attributes such as intelligence, honesty, and courage are more sensitive items than 'name, rank, and serial number.' " The researcher should determine whether his study will examine potentially embarrassing material and whether disclosure of the information could have negative repercussions for the informant (e.g., admitting illegal activity).

Sometimes it is difficult to determine the sensitivity of the information. Topics such as sexual behavior or illegal activity are usually sensitive in our society. Other topics such as religious preference and political attitudes may or may not be sensitive, depending on the group being studied. For example, Farr and Seaver (1975) found that the average college student did not consider many types of information particularly sensitive. They found that students considered *nonanonymous* intelligence tests, personality tests, and questions about drug use and family income *not* to be very private within the context of a psychological study. Even nonanonymous items about their sexual experiences were considered only a moderate invasion of privacy. Similarly, Jessor and Jessor (1975) found that high-school and college students did not mind reporting whether they were virgins. However, it is clear that many groups would consider questions about sexual behavior or family income extremely sensitive.

When the sensitivity of the information is in doubt, using pilot subjects or obtaining community input to measure the sensitivity of a proposal can help the researcher determine how private the information seems to that group. The greater the sensitivity of the information, the more safeguards (described later in this chapter) are necessary to protect the privacy of those studied. Invasions into very sensitive areas may present risks to subjects that outweigh the expected benefits of a study, and these investigations should not be conducted.

The Setting Being Observed

In 1954, law professors gained the permission of a judge and lawyers to record federal jury deliberations in Wichita, Kansas (Amrine and Sanford 1956; Burchard 1957). The jurists were unaware both of the concealed "bug" and of the project (which was designed to improve the administration of the judicial system). Like the home, the jury setting is very private in our culture, protected from outside intrusion both by laws and by tradition. When knowledge of this study became public, it created a national uproar that resulted in a congressional hearing, in insinuations that the investigators were communist subversives, and ultimately in laws banning intrusions upon the privacy of juries.

As the jury example illustrates, settings are an important determinant of privacy, and they vary along a continuum from very private (e.g., your bathroom) to completely public (e.g., a downtown sidewalk). Although the privacy of a setting may fluctuate over time (for example, one's bedroom is not private during a tour of one's home), there is generally some agreement on the privacy of a setting. The home, for example, is considered one of the most private settings in our culture; intrusions into people's homes without their consent are forbidden by statute. Other settings that are often considered private are personal offices, closed meetings, and physicians' examining rooms. On the other hand, it is usually assumed that people have relinquished the privacy of their outward behavior in settings such as sports arenas, airports, or public buses. As when obtaining sensitive information, many more precautions must be employed if behavior is to be observed in private settings. Studies of very private settings such as the home should not be conducted without informed consent.

DISSEMINATION OF INFORMATION

Imagine confiding to a close friend some very embarrassing fact about yourself and compare this with announcing the same fact over national television. Obviously there is a much greater sacrifice of privacy in the case where millions, rather than just one person, will learn your secret. The third dimension of privacy is how many people can connect personal information to the name of the person involved. Information about your sexual behavior remains relatively private if only a single scientist is informed of it. But when information is publicized, with data and names communicated through the media, privacy is destroyed. Thus, the more people who may learn of the private data in reference to particular individuals, the more concern there must be about privacy.

DIMENSIONS SUMMARIZED

Sensitivity of the information, setting, and degree of dissemination are three major dimensions that can be used to determine how private information is and what safeguards should be employed.

Cook (1976) has suggested an additional factor: whether people know they are being observed. He reasons that many people object to being watched if they have not first given their permission, even if the observations occur in a public place. Cook therefore believes that unobtrusive measures invade privacy to some extent even when carried out in public places. However, since people constantly observe each other in normal life without anyone's permission, it seems that scientists must have the same right in public places. Therefore it is probably not necessary to be concerned about the privacy of unobtrusive measures of otherwise nonprivate information.

When information can be judged as private on one of these dimensions, the problem of privacy becomes important. When two or three of the factors occur together, a very serious concern for privacy is appropriate. Consider once again the study on sororities. The information was toward the private end of at least two of the dimensions. Although confidentiality was partly maintained (real names were not published), the women were deprived of their right to privacy because a private setting was invaded without their permission and sensitive information was recorded. Since publication enabled individuals to identify each other, the safeguards for confidentiality were also insufficient. It is obvious that as one approaches the privacy ends of the three continuums, particularly when they are considered in combination, the need for safeguards becomes critical.

The greatest difficulty in the ethical analysis arises when researchers make decisions about information that is private on only one dimension or is in the middle on several of the continuums, because frequently such studies are neither clearly ethical nor clearly unethical. For example, what safeguards are necessary or sufficient in a case such as the observation of individuals in a public toilet? Mittlemist, Knowles, and Matter (1976) unobtrusively observed men in university rest rooms. They measured how long men remained at urinals when someone was standing either near them or farther away. Although the setting was public, the activity was private. Thus there is a difficult question about participants' privacy. Koocher (1977) criticized the rest room study for crossing the bounds of propriety and invading privacy. Mittlemist, Knowles, and Matter (1977) defended the study by stating that the "invasion" occurred in their subjects' lives every day, that no subject objected when he was informed of the study, and that all data were anonymous. Before we discuss ethical guidelines and necessary safeguards, several more examples of problem areas will be discussed to illustrate the issues.

Some Areas of Ethical Concern
PERSONALITY TESTING

The development of personality testing was greeted with great enthusiasm by psychologists and personnel workers alike. Tests are used now both for research and for selection for positions in schools and industry (Jackson and Messick 1967). Yet these requests for information on one's emotions, attitudes, and behaviors are inquiries into the privacy of the mind (Conrad 1967). Personality items vary in sensitivity, as the examples below illustrate. The items given here are similar to questions that appear on these tests. More complete references to tests can be found in Buros (1972, 1974) and Robinson and Shaver (1973).

1. Minnesota Multiphasic Personality Inventory (Hathaway and McKinley 1951)

   Demons control me.

   I have a hard time beginning a bowel movement.

   I find members of my own sex to be sexually inviting.

2. Fascist Scale (measuring authoritarianism—Adorno et al. 1950)

   The sex orgies of the past were mild compared to what goes on in some parts of America today.

   Criminals who sexually molest children should be flogged in public.

3. Marlowe-Crowne social desirability scale (Crowne and Marlowe 1964)

   I sometimes take advantage of people.

4. Machiavellian V Scale (Christie and Geis 1970)

   Never let people know your motives for doing something.

Many tests, as the above items illustrate, delve into private areas, and therefore research subjects should be given the option of not answering specific items. However, there are many tests in which the subject will not even know what information is being revealed; these indirect, or projective, tests do not ask for information in a straightforward way but elicit responses that the experimenter later interprets. For example, in the Thematic Apperception Test subjects are supposed to tell creative stories about ambiguous pictures. Unknown to the subjects, the stories can then be scored to indicate, for example, their need for power, affiliation, and achievement

(McClelland and Steele 1972). Just as there are indirect personality tests, so too there are disguised ways of measuring attitudes—for example, of assessing prejudice. The ethical difficulty with indirect measures is that they invade the subject's privacy without his knowledge, so that informed consent is difficult to obtain. As Cronbach (1970, p. 459), a leading authority on psychological testing, stated, "Any test is an invasion of privacy for the subject who does not wish to reveal himself to the psychologist." Indirect methods are designed precisely to circumvent participants' ability to withhold information.

One ethical concern that is unique to the testing area is the reliability and validity of the instruments. Professional standards should insist that the reliability and validity of tests be demonstrated and that test scores not be overinterpreted (Davis 1974). Further discussion on the ethics of testing is in Amrine 1965; Association of Black Psychologists 1975; Davis 1974; and Lovell 1967.

ACCESS TO PRIVATE INFORMATION
During the height of campus unrest in the 1960s, the American Council on Education sponsored a study on university protests (Walsh 1969). Thirty-five intensive interviews were conducted with students and faculty, and thousands of students completed questionnaires that assessed demographic characteristics, attitudes, and protest behavior. The data were placed in computer storage and made available to anyone who would pay the small user's fee. Since the researchers were aware of the importance of protecting their subjects' anonymity, the identification of the individuals was separated from their responses in the data bank. But there still existed the possibility of government subpoena of the material, especially since many officials wanted to put an end to student protests at that time. Thus the study of campus unrest requested private information from students but could not guarantee confidentiality in the political climate at that time. Subsequently administrators of the project made the sensitive material "subpoena proof" by storing in a foreign country the code that linked the data to individual subjects.

UNOBTRUSIVE MEASURES
The issue of privacy frequently arises when people are unobtrusively observed without their knowledge. For example, scientists conducted a study on consumer behavior in Tuscon, Arizona, by

examining garbage from different sections of the city (e.g., Rathje and Hughes 1976). They reasoned that by carefully recording the contents of garbage sacks they could discover what people actually bought and discarded, as well as what they wasted. In addition, conclusions could be drawn about waste, alcohol consumption, and so on, of cultural or social groups based on neighborhood demography. Since names were not attached to individual sacks, individual households were anonymous except where they might be identified by envelopes or letters in their garbage. The study and the many precautions taken to protect anonymity were widely discussed in the public media, but informed consent was not obtained from individual households.

A number of interesting findings emerged from the garbage study. Poorer households wasted much less food, as might be expected, than homes in higher income areas. But there was more waste of meat during the national beef "shortage" than at other times. The authors reasoned that perhaps during the meat shortage people were forced to buy cuts of meat they didn't like as well and hence wasted more. Another finding was the marked discrepancy between self-reports of consumption and the unobtrusive garbage measure. When neighborhoods were surveyed about the consumption of beer in each household, the reported rate of beer-drinking was only a fraction of the consumption indicated by the number of empty cans and bottles found in the trash. The Tucson researchers spent a great deal of time informing the community about their project and obtaining community feedback and approval before and during the study. Interviews may also be disguised as friendly interest by strangers (e.g., Perrine and Wessman 1954). Do such unobtrusive measures, with or without community approval, violate people's right to privacy? In addition, there may be a question about the legality of some unobtrusive techniques (Silverman 1975).

ANONYMITY IN PUBLICATION

A study by sociologists demonstrates the problems that may be encountered in insuring anonymity in published research. In *Small Town in Mass Society*, Vidich and Bensman (1960) described the intimate and sometimes embarrassing details of the lives of the residents of a small town in upstate New York. Although the town and its residents were given fictitious names, the individual descriptions in the book were easily recognizable as specific townspeople.

61

Vidich lived in the town of "Springdale" for two and a half years and unknown to his subjects collected observations and reported on the "Peyton-Place" activities of certain individuals. His book, which did not really insure the anonymity of his subjects, drew severe criticisms from other scientists that are published in the 1958 to 1960 volumes of *Human Organization* (e.g., Bronfenbrenner 1959). It also prompted the townspeople to stage a parade in which each wore a mask and a name tag with the fictitious names given them by Vidich, a clear indication that the whole town knew the identity of the characters in the book. At the end of the parade came a manure spreader, with an effigy of Vidich looking into the manure.

Since no consent was obtained to publish the private material Vidich had collected, protection of the townspeople's anonymity was ethically necessary. Mere pseudonyms were not sufficient for the purpose because extensive details were given about individuals. (See chapter 6 for a further discussion of the use of pseudonyms in cross-cultural research.) The authors probably could have created composite fictitious characters, although this might have detracted from the scientific merit of the publication (Vidich and Bensman 1960). It is clear that cases like "Springdale" could damage both subjects' reputations and other researchers' relations with future respondents. A researcher must think about such ethical problems before the data are gathered so that participants can be forewarned and informed consent can be obtained if anonymity of private information cannot be guaranteed. The *Anthropology Newsletter* (1977) discusses similar situations where anthropologists need to be concerned about informed consent and the anonymity of published material.

TECHNOLOGY AND PRIVACY

Only recently have concerns about privacy become acute. Before the technological developments of the nineteenth and twentieth centuries, privacy was protected by natural boundaries. Before the invention of microphones and recording devices, intrusions into the privacy of the home were impossible. But with the rapid expansion of government bureaus, national credit data banks, and research within the social sciences, concern about the vulnerability of privacy have grown rapidly as invasions of privacy have increased. There have been revelations about political information-gathering, government spying, and the misuse of credit information. With each new revelation has come a growing public concern for privacy.

Computers are a major factor in this new concern because they can store large quantities of information that can then be used by many people or agencies. It should come as no surprise that people feel their privacy is jeopardized when large computers in credit agencies can retain their complete credit history. This information can potentially be used by stores and employers anywhere in the world. However, within the realm of social research, proper precautions can actually make computer-stored information *more* private than if it were stored on paper. A large-scale project known as SCOPE, which gathered data on 90,000 students, illustrates how computer information can be stored within a large research organization to insure privacy to participants (Tillery 1967). For instance, different assistants may have access codes to different parts of the system, so that no one person can retrieve all the information. In addition, a key investigator can retain the code for each name so that respondents can never be individually identified by others. It is usually questionable to collect information for research purposes and later use the stored information to take direct actions affecting individuals, except within the context of the research project (Privacy Protection Study Commission 1977).

Electronic snooping devices may also be used to invade privacy (Schwitzgebel 1967). With such devices, conversations within a home can be heard from nearby homes. Telephoto lenses can be used to photograph events at great distances, and infrared photography can penetrate the darkest night. More conventional cameras can be used to photograph persons who might not want their pictures taken—for example, participants in a demonstration or drug addicts.

Although modern inventions make possible mammoth invasions of privacy, their use has been rare in the social sciences, and their abuse is even more rare. The most famous use of hidden devices was the "bugging" of jury deliberations described earlier (Amrine and Sanford 1956). However, even where electronic eavesdropping or recording is not clearly illegal, severe ethical problems are connected with its use to observe private behaviors without the consent of the subjects. On the other hand, when used ethically, technological innovations can benefit researcher and participant alike. For in-

63

stance, in a study of television viewing behavior, participants might be asked to record every program they viewed. Not only would this task be troublesome, but the data might be faulty because of errors, forgetting, and so forth. However, a machine can be used (with subjects' permission) that automatically records every program to which the television is tuned.

Recording and observing devices (e.g., a tape recorder or telescope) can be used ethically to record *public* behaviors—for example, patterns of window shopping on a downtown street—but often it is desirable to gain a subject's permission post hoc to use an individual photograph or a tape recording of a particular voice, even when these were made in a public place. Sometimes tape recorders and television cameras are used to record family interactions in the home. When done with the family's permission, this procedure is ethical and much less obtrusive that an observer in the home, thus increasing the probability that the family will act naturally. Similarly, a one-way mirror may assist the researcher in observing spontaneous responses within the laboratory, and there is no problem with its use if subjects know they are being viewed. But if subjects are unaware that they are being observed, each case must be examined for the extent of invasion of privacy, and precautions must be taken to protect subjects' welfare. Hence we can see that technological innovations, although they can be employed in unethical ways, can also be used ethically, making it easier for people to participate and act naturally in social science research.

Safeguarding Privacy

In the light of the high value our society places on privacy and the potential infringements on this right described above, how can the ethical investigator proceed? The very nature of the social scientist's work is observing and recording people's thoughts and behavior, which implies great potential for the invasion of privacy.

INFORMED CONSENT

Like most rights, privacy can be voluntarily relinquished (Shils 1959). A person may voluntarily surrender his privacy by either allowing an investigator access to sensitive topics and settings in his life or agreeing that the research report may identify him by name. Anthropologists regularly study the details of life in nonindustrial societies and publish information about the lives of selected individuals in those cultures. Typically, however, these social scientists have been scrupulous about gaining voluntary and informed consent from those studied. As in other areas of data collection, the type of information needed for informed consent will vary according to the nature of the research. Respondents should know how their responses will be used. If they are going to be used for a purpose they may not approve of, such as comparing their race negatively with another, they should be informed. Who has access to the data and possible secondary uses of the information may also be of concern to respondents.

Informed consent is the most unquestionably ethical way for researchers to study private areas; yet problems may arise even when consent has been obtained. For example, people may change their minds (*Anthropology Newsletter* 1976), or the information provided by a voluntary informant may violate the privacy of another person. Take the case of a man who in describing his own sex life also reveals private information about his wife's sex life and perhaps that of other women. The investigator has collected information on the sexual behavior of these other people without their consent. For instance, in studies on child-rearing practices, children could be asked to agree or disagree with questions like, "Dad always seems too busy to pal around with me." Hence, although the children are the ostensible subjects, their parents' behavior is also of interest to the investigators.

Measures of private behavior that are collected without the subjects' knowledge present an ethical problem because no informed consent is possible. Even where publication anonymity is guaranteed, the collection of the information has violated the subjects' privacy. An example of such an unobtrusive technique is a study on conversations (Henle and Hubbell 1938) in which the investigators hid under dormitory beds, listened to telephone conversations, and eavesdropped on conversations in lounges. What subjects said was recorded verbatim. Even though the results were reported only in statistical summaries without any mention of specific individuals, the observations themselves intruded upon the subjects' privacy, both in the settings involved and in the potential sensitivity of private conversations.

Perhaps prior informed consent was ethically necessary in the Henle and Hubble conversation study, but, lacking this, scientists should at least obtain permission to use such data after the fact. For example, Chambliss (1975) gained information from subjects on illegal acts they had committed, and the interviewees did not know he was tape-recording the conversations. The ethics of the study were improved considerably because Chambliss gave the tapes to the subjects at the end of the conversations and allowed them to destroy them or return them to him. Only one man (who had described a murder he had committed) chose to destroy the tape.

Many scientists will study private behavior only with participant permission; others feel that the unobtrusiveness of hidden observation is necessary to avoid measurement reactivity (Webb et al. 1966). Investigators should first be aware that invasion of certain settings such as a home or an apartment may be considered illegal trespass. If an investigator is planning an unobtrusive study that examines private behavior, he should consider the following: (1) Is there another way to collect this data? Frequently, archival records or informed consent can be used, thus circumventing ethical problems. (2) How private are the settings and how sensitive is the information to be collected? If the privacy intruded upon is minimal (e.g., sales in a store), the ethical problems are also slight. (3) Will the private data collected be absolutely anonymous? (4) Will the techniques for collecting data offend most people's standards of privacy? Will the public be upset or angry at social scientists when the data are published? It should be clear that some unobtrusive studies simply cannot be done ethically and that substantial safeguards such as informed consent are necessary whenever very private spheres are invaded.

ANONYMITY AND CONFIDENTIALITY

The willingness of subjects in past studies to reveal very sensitive information about themselves demonstrates that subjects trust experimenters to keep personal revelations confidential. Thus, anonymity and confidentiality are research tools that benefit scientists and protect subjects. Both anonymity and confidentiality require that individual identification be separated from the participants' responses. If the information is given anonymously, with the investigator unable to associate a name with the data (e.g., on a questionnaire without names), then the privacy of the informant is secure even though sensitive information may be revealed. One practical advantage of anonymity is that respondents will be less inhibited and perhaps more likely to answer honestly if their responses are unnamed (Hartnett and Seligsohn 1967; Jourard 1968). Anonymity of individual respondents in data collection is also preferable because if the data are stolen, misplaced, or perused by curious assistants, no names can be connected with the responses.

Some investigators have convinced subjects that their responses are anonymous when in fact they are not. For example, one investigator told respondents that their questionnaires were anonymous (American Psychological Association 1973), but on the envelopes in which subjects were to mail back the questionnaires, the stamps had been placed in different premeasured positions so that all respondents could be personally identified. This ruse was ethically questionable because it involved a blatant deception. Such techniques might easily destroy trust in social scientists if they became public knowledge. Since social scientists depend heavily on subjects' trust, a technique that casts doubt on their honesty is likely to be detrimental to science and should be viewed very critically.

In areas of research where anonymity is impossible, confidentiality—not disclosing the respondents' identity in any report—can still be maintained. In other words, informants' names may be attached to the data within the research organization but not published with the data or made available to outsiders. For example, the progress of schoolchildren may be followed over several years in a longitudinal study. To keep track of individuals, names must be connected to the data. But this information must remain within the confidential limits of the research team. In addition, the fewer persons within the research organization who can connect names to individual data the better, and techniques are available to minimize the identifiability of respondents by research assistants (Walsh 1969). The principles of anonymity and confidentiality should be adhered to, not only because of people's right to privacy, but also because any violation resulting in loss of public faith will be detrimental to future social science research.

Sometimes other persons or agencies request or demand information collected in a research project, but because information is confidential, these demands must not be met. For example, teachers and the school principal may be interested in data on individual children. Within an industrial setting, the management may want to see data on individual employees. In a penitentiary the prison psychologist or the warden may request information on particular prisoners. Some social scientists have been asked to donate information to computerized data banks. At times these third-party requests for information are made for punitive or regulatory reasons, and at other times the desire for information is motivated by a desire to help subjects who are in need. Whatever the reason behind the request, information should not be given to others unless subjects have given their consent. Since parents will often expect to learn of their child's performance in a study, it is usually desirable to have a clear understanding with them beforehand about whether they will have access to data on their child. Another problem that is closely related to third-party requests for information is the later use of data for a second study about which the participant has no knowledge and has not exercised informed consent. It is usually preferable to gain subjects' consent for the secondary use of their data or to keep data anonymous.

Despite the great importance of confidentiality, there are *exceptional* occasions when other concerns take precedence. For example, a researcher working in a mental hospital might discover that a patient is planning to kill a nurse with a knife he has hidden under his mattress. In such a case the value of human life would supersede the value of privacy, and the investigator would be obligated to inform the hospital authorities. Similarly, a sociologist studying an urban commando group might learn of plans to detonate a bomb in a crowded airport. Because the lives of many persons would be in jeopardy, the scientist would be obligated to disclose the plan to the appropriate authorities. When investigators discover that a person is an imminent and serious danger to society, the protection of life should take precedence over the principle of confidentiality (American Psychological Association 1973). Current laws generally support this approach. For instance, physicians, teachers, and others are now usually required to report cases of suspected child abuse (cf. Annas 1976).

Boruch (1971*a*, *b*, 1972) discusses a number of innovative methods by which anonymity and confidentiality can be guaranteed. Collected data should be safeguarded and securely stored if they contain identifiable responses. In addition, the connection between names and data should be destroyed as soon as it is no longer necessary (Ruebhausen and Brim 1966). If respondents must answer several questionnaires or participate in sessions on different dates, it is usually desirable to be able to identify the questionnaires of the same person. Code numbers or code words can be used for this, allowing the person to retain his anonymity. Astin and Boruch (1970) have described a "link" computer system for storing extremely sensitive longitudinal material. Boruch (1976) describes several methods for storing information by computer that can help insure confidentiality.

A rather clever technique known as randomized response, introduced by Warner (1965), is also designed to divorce the name of the respondent from his answer. The initial form presents the respondent with a pair of questions relating to the sensitive area. For instance, the two items might be: (1) "I have had a homosexual experience" and (2) "I have not had a homosexual experience." The respondent answers only one of the questions with a yes or no. Which question is answered is determined by a random device such as a die, used by the respondent unseen by the interviewer. For instance, the respondent privately rolls a die with the instructions that if a 1 or 2 comes up he is to answer question 1. If a 3, 4, 5, or 6 comes up he answers question 2. He then gives a yes or no response to the interviewer, who does not know which question he is answering. Because we know the probability of each question's being answered for a set of respondents (based upon the probability of the random event, such as a number's coming up on a die) the inaccuracy caused by the randomizing approach can be statistically estimated and a relatively accurate group estimate for the sensitive question derived. Some variance is introduced because the random event, for example a 1 or 2 on the die, will rarely come up the exact number of times expected by chance. However, this random error deliberately introduced to protect respondents will often be less than the error caused by respondents' lying (Horvitz, Greenberg, and Abernathy 1976). One difficulty with the randomized response technique is that some subjects may not understand what they are to do, and others may not understand it clearly enough to be convinced that they have complete anonymity.

If knowledge of illegal activities is requested, anonymity should be guaranteed so that the data cannot be subpoenaed for court proceedings or other legal matters. The social scientist's data are not automatically privileged communication under the law (Carroll 1973; Culliton 1976; L. R. Frankel 1976; Knerr 1976; Nejelski 1976; Nejelski and Finsterbusch 1973; Nejelski and Lerman 1971), and therefore congressional and court subpoenas are possible. (Currently there are several proposals for "shield law" legislation to protect the privacy of researchers' data. See Nejelski 1976; and Reiss 1976.) There are a few exceptions such as certain drug research in which statutes protect research information. If a researcher guarantees his subjects confidentiality, as is typical, then he should either make individual data anonymous or be prepared to go to jail for disobeying a subpoena (e.g., Walsh 1969). For example, during the 1950s the famous Kinsey Institute collected important new data on sexual behavior. Both the United States State Department and the FBI requested data on particular individuals who were allegedly involved in Kinsey's studies. The Kinsey Institute refused to turn over the information and announced that it would destroy its entire data files if they were subpoenaed, rather than jeopardize the anonymity of its subjects. Fortunately the government did not press the matter and the confidentiality of respondents was maintained. More recently a Harvard professor, Samuel Popkin, was imprisoned for refusing to disclose the names of persons connected with the Pentagon papers study (Carroll 1973). An alternative to making all data anonymous or refusing to honor a subpoena would be to initially warn informants that the confidentiality of the data cannot be guaranteed in case of court subpoena (Reiss 1976). Thus, if subjects divulged confidential information they would do so knowing the potential risk involved. Of course the difficulty with this last alternative is that subjects may reveal much less information if they think authorities may gain access to the information. A summary of legal and professional concerns in this area from the viewpoint of professional statisticians who use large data sets that are often collected by others (like the census and large surveys) is provided by Bryant and Hansen (1976). See also chapter 6 for a discussion of anonymity and privacy in cross-cultural and field research.

Conclusions

The cases and issues that have been presented are samples of questions related to privacy that social scientists must face. By becoming sensitive to the problems inherent in research on private behavior, the social scientist can strive to eliminate threats to subjects by providing the necessary safeguards. We suggest the following guidelines, discussed in part by Ruebhausen and Brim

(1966), in maintaining a balance between the public's right to privacy and the scientist's obligation to uncover truth.

1. As social scientists, we should be aware of people's right to privacy. We should carefully consider this right when we conduct a study, and if there are possible incursions into private areas, we should ethically evaluate the extent of invasion and possible safeguards before proceeding. When evaluating the privacy of material, a researcher should consider how sensitive it is, how private the setting is, and whether confidentiality or anonymity can be assured.

2. Whenever possible, efforts should be made to obtain the subjects' informed and voluntary consent when private information is to be gathered.

3. So far as possible, anonymity of subjects should be established. If this is impossible, all efforts should be made to insure confidentiality by making it impossible to connect the respondents with the data in published material.

4. Data that can be connected to subjects in an identifiable manner should be destroyed when no longer needed.

5. Precautions such as anonymity and informed consent should be used when research enters the most private domains. It is usually ethically unjustified to study very private areas of life without subjects' prior permission.

For a dissenting view on the importance of privacy, see Bennett 1967. For additional information, see American Psychological Association 1973; Farr and Seaver 1975; L. R. Frankel 1976; Kelman, in press; Martin and Parsons 1973; Nejelski 1976; Nejelski and Lerman 1971; *Privacy Journal*; Privacy Protection Study Commission 1977; Reiss 1973, 1976; Ruebhausen and Brim 1966; Westin 1967.

5 Deception

Sunbathers soak up the warm Florida sunshine while children build sand castles and collect shells up and down the beach. The water is warm and calm; swimmers float lazily about as a group of teen-agers splash by. Suddenly the beach erupts into pandemonium, as the cry "Shark!" sounds out. Swimmers rush madly for shore; parents scramble to grab confused children who are playing in the shallow water. Everyone who is safely on the beach stares toward the azure water, trembling, trying to spot the shark or perhaps a victim of his jaws.

Across the continent in San Francisco, a forlorn child about to cry wanders in the downtown area, looking up into the faces of adults as she goes. The urchin apparently is unaccompanied by an adult; her blond hair is ragged and dirty, her sundress torn and threadbare. Two blocks away another little girl wanders about, also apparently lost, but this child is dressed in expensive clothes. Her hair is curled as if styled at a beauty shop. An elderly gentleman stops, smiles, and gently inquires if she needs help.

What is common about these two occurrences, happening thousands of miles apart? Both are fictitious behavioral experiments, designed to confirm a theory about how adults react to and protect small children. Although these cases are fictitious, both resemble events that have actually been staged by psychologists, and both illustrate the use of research deception: the deliberate misrepresentation of a scientific study. Research deception, practiced by scientists in both laboratory and natural settings, varies along many dimensions. Some deception is by commission, involving direct lies to subjects, whereas other deception is by omission, when participants are misled by not being told relevant information. Some deceptions employ research assistants, called "confederates," who act out predetermined roles, like the little "lost" children in San

Francisco. Sometimes subjects are misled about the purpose of a study or about what will happen during the session, and sometimes they do not even know a study is in progress.

In laboratory studies, participants realize that they are subjects in an experiment but are often misled about the nature of the research. In a famous study on emotion, for example, Schachter and Singer (1962) injected subjects with adrenaline (a general arouser), telling them the drug was a vision-enhancing chemical to be used for a study on vision. After the injection, each subject was placed in a waiting room with a confederate who acted either euphoric or angry. Subjects who were unknowingly aroused by the drug tended to think their mood was similar to the confederate's. The experimenters were interested in how subjects would react and what emotions they would feel. Note that in this study several deceptions were employed: the contents of the injection were misrepresented, there was a confederate, and the study's true purpose was concealed.

Outside the laboratory a common research deception is the simulated emergency requiring the intervention of a passerby. Thefts, medical emergencies, and searching for "lost" contact lenses have been faked by confederates, enabling scientists to measure people's reactions (e.g., Bickman and Henchy 1972; Gaertner and Bickman 1971; Swingle 1973). Latané and Darley (1970), for example, faked a theft of a case of beer from a liquor store, to see whether patrons would be more likely to report the theft to the manager when they were alone or when the store was crowded. Of course the store owner was "in" on the study. When the beer theft study was replicated in the Midwest by other investigators, one customer called the police from a public telephone near the store. The officers, unaware of the study, arrived with guns drawn to apprehend the frightened researchers—one of the few deception studies in which the researchers were more embarrassed than the participants.

Deception in field settings has also been employed to gain access to observations that might have been closed to the investigators. A sociologist, Albert Reiss (1968, 1971), was interested in how police treat citizens. Charges of police brutality were frequent at the time; yet these charges had never been verified by trained and objective observers. Reiss knew that had the true purpose of his study been known to the individual policemen who were being observed, it would probably have drastically curtailed police brutality. He therefore led the officers to believe that the study mainly concerned citizens' reactions to the police. He justified the ethical cost of deception by referring to the public's right to know about the behavior of public employees. The observers recorded an amazing amount of police mistreatment and even brutality in the course of the study. Despite the important findings, the ethical dilemma was that the individual police officers did not exercise fully informed consent, since they did not know the purpose of the study or realize that they were the primary subjects. However, the informed consent of the police chiefs had been obtained, and the anonymity of the officers was protected. The study clearly reveals how deception may be used to gain access to information that might otherwise be closed to the investigator. In addition to the ethical issues of research deception and informed consent in the Reiss study, another serious potential problem is that research of this type might generate distrust so that future researchers will find it difficult to gain the cooperation of participants such as police officers. Reiss has been highly involved in the discussion of ethical issues related to research, and undoubtedly he weighed these considerations carefully before beginning the study.

Research employing deception has become commonplace. Deceptions like those described above are not techniques used only to study important questions after exhaustive consideration of alternate methods. Several reviews of the published research in psychology (R.J. Menges 1973; Seeman 1969; Stricker 1967) suggest that between 19 and 44 percent of recent personality and social psychological research has included direct lying to subjects. Menges (1973) estimated that complete information was given in only 3 percent of the psychological studies he reviewed.

Although research deception is widespread, the ethical problems of particular deception studies vary greatly, depending on the nature of the deception employed. For example, dropping "lost letters" to see how many will be returned (Milgram, Mann, and Harter 1965) represents a minor deception. It entails no risk to participants, the deception is by omission (subjects were not told that it was not a "real" letter), and the magnitude of the deception is small. Contrast this to the case presented earlier (see chapter 3) in which volunteer patients were injected with a muscle-paralyzing drug (Campbell,

74

Sanderson, and Laverty 1964) that they were told was a possible cure for alcoholism. In this study there was a potential for physical harm—even a small chance of death—and extreme fright was almost certain. Recognizing that the ethics of the two cases of deception differed widely, the reader might nevertheless ask: Since lying is generally considered immoral in our society, why carry out a study that employs deception of any type? As we shall see, although serious ethical concerns are raised by research deception, there are also a variety of reasons for its use.

Reasons for the Use of Deception

Deception has been used because it offers a number of methodological and practical advantages over other methods. The arguments for deception fall into several important areas: methodological control, external validity, pragmatic reasons, ethical considerations, and possible lack of negative effects. These arguments, taken together, represent a plausible defense for the use of some deception research.

METHODOLOGICAL CONTROL

Deception has sometimes been employed to create a "fictional environment" (Seeman 1969), so that a specific variable or situation can be clearly manipulated and controlled. In conformity studies, for example, investigators have been interested in how a participant will respond when a group (all confederates) makes unanimous judgments that seem to contradict normal perception. The confederates may all insist, for instance, that the shorter of two lines is the longer one. Since it would be impossible to find an authentic situation in which all other participants would disagree with a subject about such an objective matter, deception was employed to create an extreme amount of conformity pressure. In addition to creating situations that may otherwise be unavailable for study, deception allows closer control of variables so that their precise effects may be studied. For instance, in conformity research the amount of conformity pressure can be manipulated by varying the number of confederates who agree to a judgment that contradicts the subject's perceptions or beliefs. Fictional environments allow a great deal of control over variables so that their effects can be assessed independently of other influences. For many situations, this

75

kind of control would be difficult or impossible to obtain otherwise.

EXTERNAL VALIDITY

External or ecological validity is the extent to which findings can be generalized from the experimental situation to natural settings in which the scientist is interested. Even in the laboratory, deceptions are usually designed to elicit spontaneous behavior and thus make the findings more generalizable to everyday behavior (Aronson and Carlsmith 1968). Within behavioral science, the act of measuring something can often change what is measured (called "reactive measurement"), because people often alter their behavior if they know it is being observed and recorded (Webb et al. 1966). In both the laboratory and the field, deception is often practiced to reduce reactive measurement effects and thus increase the external validity of the findings. For example, if people know the researcher is measuring some socially undesirable trait such as aggressiveness or conformity, they may avoid the behavior so they will not be judged negatively. In surveys, interviews, and observational studies, subjects are often not told the purpose of the questions or observations, and sometimes are not even told they are being observed, so that they will answer or behave spontaneously.

In field situations, people also might avoid the situation entirely or act unnaturally if they know they are being observed. Consider the study conducted by Gaertner and Bickman (1971) in which people were phoned and requested to help a "stranded motorist" who had dialed their number by mistake. The caller claimed to have used his last dime and asked the person answering to telephone a garage for him. If subjects had realized it was an experiment, they probably would have either phoned so as to appear helpful to the researchers or not phoned at all, realizing that the caller was in no real need of help. In either case they would not have been acting naturally and altruistic behavior would not really have been the subject of study. In sum, fictional environments are designed to elicit spontaneous behavior, thereby increasing the generalizability of the findings to natural situations.

PRAGMATIC REASONS

Practical considerations such as limited time, limited money, and the inaccessibility of private observations have often encouraged the use of deception. For example, one might collect data on how people react to natural emergencies by sending hundreds of research assistants to busy locations throughout London for a month and waiting for real emergencies to occur; but many times this would be impractical, since these natural events may be rare. Therefore a practical alternative is to stage the events and record people's reactions to the simulations.

Another practical consideration concerns the possibility that some persons or groups may not consent to be observed unless investigators misrepresent themselves (see also chapter 7). Festinger, Riecken, and Schachter (1956) wanted to test cognitive dissonance theory in a realistic setting and found for their purpose a small religious group that believed the world would soon end. The scientists reasoned that when the predicted doomsday failed to occur, dissonance would develop in the believers. The researchers' purpose was to discover how the dissonance would be resolved. The psychologists joined the group as "faithful believers" and did not reveal their identity as social scientists. As we know, the world did not come to an end, placing the true believers in a dilemma. Some resolved their dissonance by deserting the religion. Others, notably those who had been with their leader on the predicted night of destruction, became even more faithful. Their leader brought them a message from outer space that the world had been saved because of the faith of this small band. These members then worked hard to convert others. It is unlikely that the scientists would have been admitted to the group had their true identity and lack of faith been revealed. In addition, Reiss's study of police brutality mentioned earlier also illustrates the possibility that, even when subjects know they are being observed by scientists, they are more likely in some cases to alter their behavior if they know the true nature of the study. Just as Festinger et al. used concealment to gain their subjects' cooperation, some laboratory investigators insure that volunteers remain in the study by misrepresenting parts of the research to which some subjects might object.

ETHICAL CONSIDERATIONS

As contradictory as it may at first appear, deception is sometimes practiced to make a study more ethical. For example, the deception practiced in aggression experiments usually consists of telling subjects they will be administering painful electric shocks to another subject. Actually, the apparatus is constructed so that the confederate receives no shock. It would be possible to allow the subject to really shock another subject or a confederate, but the ethical problems in having a second person needlessly undergo pain are greater than those involved with deception. Similarly, stress has been induced by telling subjects they are about to receive a painful shock. Once the stress reaction has been recorded, however, the investigator would not want to deliver the shock if it were not a necessary part of the experiment, simply to make his earlier statements true. In other words, lies are sometimes employed to spare subjects from stress or possible harm. But this point should not obscure the fact that the deception itself may be harmful and that at times it might be preferable to cancel the entire study rather than use deception for questionable purposes.

POSSIBLE LACK OF NEGATIVE EFFECTS

Some suggest that the negative effects of research deception may not be serious (Elms 1977). There are data that indicate that people do not seriously object to deception when it is carefully restricted to research settings and is necessary to the study. In a study conducted by Epstein, Suedfeld, and Silverstein (1973), it was discovered that the typical research population—college students—generally did not consider research deception evil. The majority expected to be deceived as participants in psychological research, and this was permissible from their viewpoint. Although the students considered being deceived in research personally undesirable, they did not think it was inappropriate. Rugg (1975) found that other groups also did not strongly object to deception. He queried a variety of psychologists, college professors, lawyers, and students and found that most of them did not object to deception per se. Rugg, as well as Sullivan and Deiker (1973), found that college students are generally less strict about the ethicality of most experimental procedures than are psychologists, even though students are the group most frequently used as subjects and the ones most likely to suffer any harm resulting from deception. Wilson and Donnerstein (1976) found, however, that a small percentage of a sample of adults who were surveyed in public places did find some of the deceptions objectionable.

Even stronger evidence for the lack of negative effects of deception comes from studies where deception has actually been employed. MacKinney (1955) found that students who had been deceived in a classroom experiment (they were falsely told that they could choose the type of course examination they would have) were not disturbed about the deception afterward. MacKinney hypothesized that "being deceived as an experimental subject is more a problem when viewed in the abstract than when it actually happens to the individuals concerned" (p. 133). In another study, Ring, Wallston, and Corey (1970) interviewed subjects after a Milgram-like obedience study in which subjects were ordered to give very high voltage shocks to another "subject" who was actually a confederate. The authors reported that debriefed subjects did not resent the deception or regret participating in the study, and most thought of participation as a positive experience. None of those deceived thought that the study was unethical or should be discontinued. Similarly, Clark and Word (1974) found that 94 percent of their subjects viewed deception as unavoidable in the emergency intervention study in which they had just participated. Further, when questioned several months later, 92 percent of the subjects did not feel that their rights had been violated or that they would rather have avoided the experience.

SUMMARY OF THE CASE FOR DECEPTION
The case for research deception can be summarized as follows: Because of practical, methodological, and moral considerations, much research would be difficult or impossible to carry out without deception. Important knowledge gained in the past would have been forfeited had the practice been totally abandoned. There is no evidence that anyone has been harmed by it, and students do not seem to object strongly to being deceived. As long as deception is practiced within the well-understood and circumscribed limits of research and no one is harmed, it is not unethical, and in fact it has been employed in some outstanding studies.

Problems with Deception
Despite the seemingly compelling reasons for the use of deception in research, the technique has been widely attacked on both moral and methodological grounds. Some critics attack any use of deception (Shils 1959; Baumrind 1964; Seeman 1969), whereas other critics, while not condemning all deception, nevertheless maintain that it has been overused and frequently employed carelessly (e.g., Kelman 1967; Ring 1967). The arguments may convince some that the technique is never justified. For others who decide that research deception is not universally wrong, the criticisms of deception

should help clarify when the method might be justifiable and what precautions are necessary to minimize potential negative effects. The problems with deception may be divided into several interrelated areas: suspicion of subjects, further methodological problems with laboratory deception, ethical concerns, societal effects, harm to the scientist, and limiting informed consent.

SUSPICION OF SUBJECTS

Although the strongest arguments supporting most uses of deception are methodological, there is evidence that deception itself causes a new set of methodological problems. The most serious of these arises when subjects' knowledge and suspicions about experiments cause them to act unnaturally in studies (McGuire 1969; Stricker, Messick, and Jackson 1969). Many subjects have become sophisticated about research deception from college courses, friends, previous research experience, and even the public media. Subjects' sophistication about a study ranges from vague knowledge that research deception occurs to specific foreknowledge of the particular experiment. When participants suspect they are being hoodwinked, they may sabotage the study. For example, Argyris (1968) found students in a subject pool expressing resentment and a desire to beat the experimenter at his own game. Sidney Jourard (1968) summed up the potentially negative effects of the growing distrust and cynicism in an imaginary letter from a subject to an experimenter. The letter exemplified the attitudes and feelings subjects had communicated to Jourard about research experiences:

> You often lie to me about your purpose.... It's getting so I find it difficult to trust you. I'm beginning to see you as a trickster, a manipulator. I don't like it. In fact, I lie to you a lot of the time, even on anonymous questionnaires. When I don't lie, I will sometimes just answer at random.... Then, too, I can often figure out just what it is you are trying to do, what you'd like me to say or do; at those times, I decide to go along with your wishes if I like you, or foul you up if I don't.

Rates of Suspicion. Although early reports of the amount of suspicion among research subjects were reassuringly small (Stricker 1967), several recent investigators have found larger percentages of suspicious subjects (Gallo, Smith, and Mumford 1973; Stricker, Messick, and Jackson 1967; Glinski, Glinski, and Slatin 1970). They discovered that between 50 percent and almost 90 percent of participants voiced some suspicions when inquiries were made after the session. Stang (unpublished) has found that in published research the amount of reported suspicion and the year showed a positive correlation of .76, indicating a steady rise in suspicion during the last two decades. When one examines the increasing lack of naiveté among research subjects, the obvious question arises: Who is really being deceived? (Brown 1965; Kelman 1967; Newberry 1973).

As knowledge of deception has increased among college students, investigators have turned to other subject populations. There is evidence now, however, that even high-school students are highly suspicious (Stricker, Messick, and Jackson 1967). It is only a matter of time before each new population becomes wise to the ways of social scientists. New settings outside the laboratory have also been utilized in hopes of avoiding suspicion. But as these studies are discussed in classes and described in magazine and newspaper articles, people may become suspicious even of everyday occurrences. Deception can be likened to the use of DDT. At first it is effective, but people soon become immune to it. One must then step up the rate of deception or seek out new people and places. Soon high rates of suspicion will be found among the new subject populations. Eventually deception may only occasionally be effective, and in the meantime the environment will have been polluted by its use (Bonacich 1970).

Effects of Suspiciousness. Subjects' suspicions may alter their reactions enough to make the results of the experiment misleading. Participants may try to act as they think the experimenter wishes them to (Orne 1962; Fillenbaum 1966), or they may try to behave in a way opposite to what they believe is the experimenter's hypothesis because they resent the deception (Masling 1966). In either case the results will not be valid. There is now evidence that suspicious and preinformed subjects do produce results unlike those of naive subjects (Allen 1966; Golding and Lichtenstein 1970; Levy 1967; Newberry 1973; Stang, unpublished; Stricker, Messick, and Jackson 1967). This indicates that prior deception or suspicion frequently may have a confounding effect on later research. And this confounding, as we shall see, is largely undetectable.

Reports of research in which suspicious and reportedly unsuspicious subjects behave the same are not reassuring. First, there is a strong probability that many suspicious subjects have not been identified. Second, there is a problem in interpreting findings when suspicious and unsuspicious subjects behave identically. Remember that subjects in deception studies are supposed to be reacting to what they believe is a real situation. However, for suspicious subjects the deception has not been entirely effective. When subjects who suspect the fictional environment act exactly like those who do not, one can no longer be certain about the true cause of the results. The possibility exists that the behavior of both suspicious and unsuspicious subjects has been caused by some extraneous factor, such as demand characteristics of the experimenter, that is not under study.

An even more serious problem with suspicion is that it may often exist differentially between various conditions (Stricker, Messick, and Jackson 1967; Stang, unpublished); that is, different treatments are likely to create different levels of suspicion. Suspicion rates may thus account for apparent treatment effects in some studies. Stang (1976) found that the rate of suspicion was not randomly distributed across conditions, and omitting suspicious subjects from the analysis increased the ostensible treatment effects. On the other hand, the experimenter who discards self-identified suspicious subjects loses the truly random assignment of subjects and has no way of knowing if treatment effects are due to subject selection biases. Nor can one ever know whether undetected suspicion exists differentially across conditions and is responsible for the apparent treatment effects.

Knowledge about deception can change the behavior of subjects, whether or not deception is involved in the particular study. There is evidence that those who have participated in deception research often behave differently in subsequent studies (Brock and Becker 1966; Cook et al. 1970; Fillenbaum 1966; Silverman, Shulman, and Wiesenthal 1970; Stricker, Messick, and Jackson 1967). Thus the hapless experimenter who is telling the truth may fall victim to subject suspicion created by his less truthful colleagues. Students often approach studies as if watching a mystery movie: everyone and everything is suspect. They may not know just what the experimenter is up to, but their general suspiciousness may still change their behavior. Participants may "play" the experiment as a big game; they may try to make themselves appear bright and socially adept or be very cautious so they are not made fools of. If they have built up resentments from previous experiments, they may even try to sabotage the study. The analogous shortcoming in field research is that subject groups may not trust a scientist if they have previously been duped by other researchers or heard about the dishonesty of behavioral scientists.

Leakage of Research Information to Future Subjects. Even more distressing than general suspicion is the case where subjects learn details of the specific study from previous subjects. Explicit knowledge of the true purpose of the study or other concealed facts will probably be more detrimental to research results than vague suspicions would be. Some researchers have therefore taken the precaution of either never revealing the deception at all or not revealing it until the research has been completed. However, these expedients postpone debriefing for some time, allowing potential adverse effects to go unchecked.

Research to date suggests that "leakage" of experimental information may vary greatly, depending on the nature of the experiment and, perhaps more important, on the nature of the subject pool. A number of investigators have found evidence for only minimal amounts of information transmission into the subject pool (Aronson 1966; Diener, Matthews, and Smith 1972; Lichtenstein 1970; Stricker, Messick, and Jackson 1967; Wuebben 1974). Others have found evidence suggesting leakage to an extent that could substantially affect the outcome of the experiment (Glinski, Glinski, and Slatin 1970; Wuebben 1967). Walsh and Stillman (1974) found evidence supporting the hypothesis that subjects making an oral or written agreement to keep silent were much less likely to divulge information to others (20 percent did so) than were those who had not made the agreement (80 percent divulged information), suggesting that contracts of silence are a useful expedient. Rokeach, Zemach, and Norrell (1966) suggest that researchers use order of collection as one factor in the data analysis to check on possible information contamination effects. Leakage is simply another possible source of confounding to be added to widespread suspicion.

Detecting Nonnaiveté. Although subjects who are suspicious or who have received prior information about the study pose potential methodological problems, these could be partly corrected if such persons could be identified. We now know, however, that most preinformed and suspicious subjects are unlikely to reveal their doubts to the experimenter. An experiment by Golding and Lichtenstein (1970) demonstrated that most subjects will not admit suspicion to the experimenter even if they have previous knowledge of the study. In this study subjects heard over a loudspeaker a fake heartbeat that was supposedly their own. The rate increased as the subjects looked at pictures of *Playboy* magazine nudes. Before the experiment began, a confederate tipped off one group of subjects about the phoniness of the heartbeat, so that they definitely knew of the experimental deception. Yet in a debriefing in which "scientific integrity" was stressed, fewer than half the subjects confessed full awareness of the deception. Levy (1967) similarly used a confederate to give subjects precise information about a deception experiment before it began. In a postexperimental interview, only one of sixteen subjects admitted full previous knowledge of the research, and only 25 percent admitted having any prior information whatsoever. Denner (1967), Newberry (1973), and Lichtenstein (1970) also gathered evidence indicating that most subjects do not admit suspicion. These findings are disturbing because they suggest that even careful questioning will probably not reveal which subjects are suspicious, and therefore reported research data will probably include responses from a number of suspicious persons.

Most researchers are not anxious to uncover suspicion that can negate their research effort. Many will look quickly, if at all, for suspicion, and it is unlikely they will find it. This conspiracy of silence (Orne 1962) allows both subjects and experimenters to naively assume that they are advancing scientific knowledge together, whereas in fact the attempted experimental manipulation might be a total failure. One can see that the methodological shortcomings accompanying deception research are substantial. Thus far these problems have not been adequately solved.

FURTHER METHODOLOGICAL PROBLEMS WITH LABORATORY DECEPTION

Sometimes laboratory researchers attempt to minimize the problems of external validity by employing situations that on the surface look like natural situations. But in deception research in the laboratory, behavior that is outwardly similar to some form of "real world" behavior may be defined quite differently by the subject (Forward, Canter, and Kirsch 1976) and therefore is not a true equivalent. In laboratory studies of aggression, for example, administration of shock to an experimental confederate is often disguised as part of a study on punishment and learning. Since a noxious stimulus is supposedly delivered to another, the "aggression" is superficially similar to real-world violence. But the meaning of the behavior may be quite different. The experimenter's attempt to camouflage the behavior so that it does not seem like aggression may destroy its usefulness as a real-world analogy because of the change in its subjective meaning (Baron and Eggleston 1972).

Deception is used to simulate everyday situations, yet evidence is beginning to accumulate that laboratory behavior often cannot be generalized—that is, it does not have ecological validity. College students do not throw acid in other people's faces just because they are asked to do so, as they have done in the laboratory (Orne and Evans 1965). Most people will not electrocute others in everyday life just to teach them some word associations (Milgram 1963). That people readily do these and other extreme acts in the laboratory is testimony that an experiment is different from everyday life. Whereas the problems of external validity apply to all laboratory studies, deceptive and nondeceptive alike, these arguments reveal that deception does not automatically bestow external validity upon a laboratory study.

Another methodological shortcoming of deception in the laboratory is that it limits the range over which variables can be studied. It is well known among investigators that just the right strength of a manipulation must be used. If it is too strong, suspicions will be increased and there will be no results. If it is too weak, the manipulation will have no effect. Consider for example, a case where an experimenter wants to anger subjects in an aggression study. If subjects are at all sophisticated, they will quickly become suspicious when insulted directly. If there is an insulting confederate, subjects nowadays often immediately suspect that he is working for the investigator—else why would he act in such an atypical way? So the experimenter may compromise by making the insult indirect or subtle. But then some subjects will not be angered at all and others will only become resentful. Note that anger can be studied only within a rather narrow range and that its effects might be quite different at other levels of intensity.

ETHICAL CONCERNS
The earliest and most basic argument against deception is that deception is basically unethical and degrading (Kelman 1967;

Seeman 1969; Vinacke 1954; Warwick 1975). Almost all moral systems consider lying to be evil, and major religions such as Buddhism, Christianity, Judaism, Confucianism, and Hinduism all teach that falsehoods are wrong, perhaps because they destroy the trust upon which good human relations are built. Such arguments hold that it is irrelevant that subjects do not protest against deception, because the practice is intrinsically wrong. People are interdependent and need each other for survival and for psychological health. Lying in a research context becomes doubly troublesome when one realizes that the ultimate value of science is "truth."

Many people have justified deception in terms of important ends. President Nixon tried to justify lying to cover up Watergate by referring to the value of a strong presidency and international respect. The Central Intelligence Agency of the United States justifies deceiving the nation by insisting on the need for national security. A business executive rationalizes corporate lying by talking about the importance of a strong economy. Wrongdoing is often explained by reference to some "higher" value, but this is usually no more than a rationalization for the selfish interests of a particular person or group. Since falsehood can destroy the trust, interdependence, and mutual respect upon which society is built, the critics of deception maintain that scientists should adhere to honesty for the benefit of society.

SOCIETAL EFFECTS

Closely related to the ethical problems of deception just discussed are the destructive effects experimental lying may have on society. Students may see high-status professors encouraging the use of extensive deception. In this way deception may be adding to a cynical trend toward treating other people as objects to be used and manipulated. Deceptions that endanger traditional role relationships have very strong negative effects on society. For instance, in one study discussed by Seeman (1969), psychiatric nurses gave false feedback to hospitalized mental patients. The nurses, as part of a study, told the patients they were in serious condition. This deception could have broken down patients' trust in the nurses, a trust usually considered essential in therapeutic settings. The deception was doubly destructive because giving already disturbed persons negative feedback was probably antitherapeutic. A similar case was a study on cheating among college students by Smith, Wheeler, and

Diener (1975). In this study the instructor gave students an easy opportunity to cheat on the midterm exam. However, there was an unobtrusive way to determine which students cheated. This study endangered a trusting professor-student relationship and could also have contaminated fair grading practices. The deception was especially undesirable because it encouraged students to transgress societal norms.

When deception studies are moved to field settings they may have another negative consequence; people may come to think of real situations as "just psychology experiments." At the University of Washington in Seattle in 1973, a male student accosted another student on campus and shot him. Students on their way to class did not stop to aid the victim, nor did anyone follow the assailant (who was caught anyway). When the campus reporters asked some students about their lack of concern over the murder, they said they thought it was just a psychology experiment (Gay 1973). This and similar incidents around the country raise the possibility that people familiar with social science research methods may be less trusting and act less naturally in their everyday life because of their belief that a real-life incident may be only an experiment. Such occurrences demonstrate one dramatic negative effect that social science research can have on society (see chapters 6 and 7 for further discussion of deception in field settings).

HARM TO THE SCIENTIST

Another reason critics maintain that social scientists should avoid deception is the possible negative effects it may have on the scientists themselves (Mead 1961). First, deception may lead the scientist to think of people as objects to be manipulated. Second, researchers may later suffer guilt over lying. In a previously cited study (chapter 2), it was noted that a student, Gerry Davison, acted as a confederate who asked coeds for dates (Walster 1965). When approached later, Walster admitted to being disturbed at having to tell coeds that their dates were canceled and Davison said he regretted ever participating in the study and questioned whether the research was really valuable enough to justify the deception involved (Rubin 1970).

Reliance on deception could even make a person a poorer scientist. Ring (1967) has argued that a fascination with deception has led experimenters to adopt a gamelike approach to research.

They begin to have so much fun at the challenge of tricking subjects that they get caught up in clever and flashy studies rather than tight, rigorous research. In a research atmosphere of "See if you can top this," clever deceptions are often of more concern than discovering knowledge. With deception enshrined as *the* paradigm in some areas such as social psychology, students may not seek out new methods or see the limits of the old. They may develop into researchers who take up only problems amenable to the deception paradigm.

LIMITING INFORMED CONSENT

Not only is deception itself an ethical problem, but its use may limit informed consent. When faked events are staged in natural settings, subjects are not given an opportunity to volunteer or to refuse to participate. They become participants without their knowledge or consent. Similarly, deception in the laboratory makes informed consent more difficult to obtain because the subject does not know the true nature of the study. Where informed consent is ethically necessary, deception is ethically unacceptable if it interferes with the subject's understanding of important facts about the study that may influence his decision to participate.

Summary and Resolution

The arguments both for and against deception are persuasive, and we will not offer a definite conclusion favoring one side of the debate over the other. However, we can point out several factors readers should consider in deciding the issue for themselves. Several of the arguments for and against deception are methodological and must be resolved in different ways for different cases, through further research and individual investigators' experiences. Some methodological problems can be overcome by careful planning and implementation of the study (Wuebben 1974). The limitations of the technique are becoming progressively clearer, and we hope the overuse of deception will consequently decline. Subject suspicion is currently a major shortcoming of the technique, especially with the growing knowledge of deception among college students. There are alternative methods, some of which are more suitable for certain research purposes, and these should enjoy more widespread use in the future.

The most telling criticism of deceptive research is that lying is immoral. Although most investigators would generally condemn lying as wrong, in candid moments many would state that they see nothing wrong with lies that don't harm anyone. Nevertheless, one should remember that interpersonal and societal trust, as well as science itself, is built upon the value of truth. The reader must personally decide the issue of the morality of research deception. However, many may not decide the issue in an absolute way but may conclude that whether deception is justified will depend on the particular case.

CONSIDERATIONS FOR JUDGING DECEPTION STUDIES

When considering the question "Is deception research ethical?" one should remember that deceptions are not all the same, and ethical judgments about the technique vary greatly depending on its specific use in a study. The investigator who thinks of conducting a deception study may judge it in reference to three factors: the need for informed consent because of potential harm, the magnitude of the falsehood, and the degree to which deception is restricted to a traditional research setting.

Informed consent. The greatest ethical problem with research deception occurs when subjects are exposed to risks they do not know they are undertaking. When the rights relinquished or the risks undertaken are substantial, informed consent is necessary. Deception may be employed in studies where informed consent is necessary, but the subject must always be given enough knowledge of the probability and magnitude of risks to make an informed and free choice whether to participate.

Magnitude of falsehood. It is obvious that the degree of falsehood may vary greatly, from extremely small to very large. In minor deceptions subjects might be deceived about one detail of a study or might be led to believe that their data will be used even though they are only participating in a pilot session. In a large-scale deception, the subjects might be misled about the entire purpose and procedures of the study. Where the deception is by omission or is relatively small—for example, dropping "lost letters" or leaving headlights on in a car—few would consider it unethical. But where falsehoods are either directly stated or substantial, a careful ethical analysis is necessary before the study is carried out.

Deception circumscribed. Other things being equal, deception within a known research setting such as a laboratory is considered more ethical. This judgment centers on the idea that if deception can be restricted to a clearly specifiable research situation, it is less likely to have destructive effects. Since the public is becoming aware of research deception, it is preferable that they see it as justifiable only within a very limited context. Thus, research deception in the laboratory, like deception in the theater, will be seen as a special case that cannot be used to rationalize falsehood in other areas of life.

Each of the three factors above should be helpful in judging potential deception studies. Cost/benefit analysis was suggested in chapter 2 as a helpful initial approach to ethical decisions. The three criteria outlined here should help in weighing the costs of the deception. In addition, the more extreme the research deception, the greater the safeguards required to protect subjects.

Safeguards

If an investigator decides that a particular study employing deception is ethically defensible and that he is personally willing to accept the moral costs involved, precautions must be taken to minimize potential negative effects. As stated above, informed consent is a necessity in cases where substantial risks are taken or rights surrendered by subjects. In addition, the experimenter should use the minimum of deception. Usually it is ethically preferable to omit information from the instructions rather than to lie directly. For example, it would be ethically superior to omit mention of the purpose of the study altogether rather than to make up some fictional purpose. In addition to these considerations, there are several safeguards that should help protect subjects: using subjects in only one deception study, peer review, and debriefing. Using each participant from a subject pool in only one deception study should help minimize negative effects. If subjects are deceived repeatedly, they are more likely to become hostile or cynical toward social science. Subjects' sense of interpersonal trust may be left intact after one deception, but this will become progressively less likely if they are repeatedly deceived. This precaution will also benefit the researcher, since those who have participated in previous deception studies may behave atypically (e.g., Cook et al. 1970). Another safeguard that will help insure that deceptions are not too objectionable to subjects is to have a local committee of professors and a panel representing the potential subjects of the study review the procedures ahead of time.

DEBRIEFING

A major safeguard that is often used with laboratory deception is complete "debriefing." The term refers to the experimenter-subject interchange at the end of the session. During this time subjects are supposed to be told the true nature of the study and be educated about its importance, have questions answered, and be reassured if they feel upset. Debriefing, which should usually be used in both deceptive and nondeceptive research, should not only try to assure that subjects are not upset and feel positive about their experience, but also be educational (Tesch 1977). Subjects should be told how they can get a final report of the results of the investigation. Debriefing should be conducted in accordance with the dictum that subjects should not leave the experiment with less self-esteem or more anxiety than they entered with (Baumrind 1964). In fact, researchers should try to follow the advice of Kelman (1967), who said that since participants are giving us something, we should leave them better off than before—possibly with more self-esteem, or less anxiety, or with greater self-understanding. Evidence shows that debriefing can make experimental participation a positive experience (Gruder, Stumpfhauser, and Wyer 1977). Judson Mills (1976) has reported on a complete debriefing procedure that he has developed over a twenty-year period. Mills's article contains an excellent model of a debriefing protocol. Holmes (1976*a, b*) has reviewed evidence suggesting that debriefing is effective both in giving subjects an accurate picture of what occurred and in helping them deal with any new information about themselves they acquire from their behavior in the experiment.

A positive payoff for subjects can often best be guaranteed by a careful debriefing at the end of the research. But if subjects have been selected because of a somewhat undesirable trait such as low self-esteem, the investigator will probably leave the subject in better condition if this is omitted from the debriefing. At the conclusion of the research, a careful assessment should be made of each individual and any resultant anxieties should be eased. Subjects should be encouraged to reveal possible negative effects and be helped to work through their feelings. If there is a potential for serious upset, arrangements should be made for a "backup" such as a clinical psychologist to aid any disturbed subjects. In research that is apt to be especially stressful, a follow-up check should be made at a later date. The necessity for the deception should be explained so that it does not appear arbitrary or capricious. The participants must understand that the researcher values their participation and wants them to know the exact nature of the experiment.

Since one of the subjects' major gains from research participation is the educational experience, they should be informed about the experiment at their level of understanding. MacKinney (1955), for example, has found that most subjects want a full explanation of the purposes of the study afterward. The justification for most university subject pools is that they are an educational opportunity (see chapter 11), but the extent to which this is true depends on the investigator. Time should be allowed after experiments for this educational function, which is an especially serious obligation when participants are not paid.

One way the debriefing can add to subjects' self-esteem and their liking for future research is by stressing the important contribution the subject has made to scientific knowledge. Subjects should not be treated as just pieces of data. Instead, it should be made clear to each that his individual participation has been helpful. The subject should be treated with respect and sensitivity, which of course are desirable throughout the experimental session. After the study, an assistant whose sole task is debriefing can often talk to subjects in a friendly, conversational way. Such treatment will help convince the subject that he has personally made a worthwhile contribution to science. A thorough debriefing includes education, careful relief of anxiety, and helping the individual feel he has made a worthwhile contribution to science. This requires a large time commitment by the experimenter, and some may therefore continue to rationalize away the need for debriefing and give out a dull mimeographed explanation or a perfunctory thirty-second oral debriefing. Yet those who maintain that deception is indispensable to the advancement of knowledge in social science should be willing to bear the costs of minimizing the harm of the technique.

Although debriefing is an important safeguard, it cannot be used indiscriminately to justify all research deception. Debriefing is rarely employed in field studies, but even in laboratory studies where it is used evidence indicates that it may not totally eliminate experimental effects. Ring, Wallston, and Corey (1970) conducted an obedience-shock study and used three different debriefing conditions for different groups of subjects. They found that participants who were not debriefed about the obedience-shock study were consistently more negative about the experiment later, and that participants who were given a justification for why they had behaved the way they did evinced little tension later. Subjects who were given a debriefing that implied blame for their behavior experienced more negative affect than those in the justification-debriefing group. Regardless of the type of debriefing, Ring, Wallston, and Corey (1970) found that a few subjects continued to be upset afterward even though they said the experiment was a positive experience. These results suggest that a carefully planned debriefing may help alleviate anxiety but may not be entirely successful in returning subjects to their previous state.

Further evidence on whether debriefing completely eliminates experimental effects reveals that although it often does so, sometimes it does not (Abrahams 1967; Holmes 1976a; Holmes and Bennett 1974; Ross, Lepper, and Hubbard 1975; Walster et al. 1967). A particularly instructive investigation on the effects of debriefing was employed in an altruism study conducted by Hammersla (1974). Some subjects played the role of persons in dire need of medical help, pretending to be suffering from epileptic attacks. Another subject had to walk past the room where the "patient" lay gagging on the floor; some of the subjects helped the victim and others did not. Both groups of subjects were then debriefed about this phase of the research. Several weeks later these subjects were individually confronted with another helping situation outdoors, blocks from where the first study occurred. A young college girl, crying and noticeably disturbed, dropped all her belongings a few feet from where the subject was walking. The percentage of persons helping her varied according to what group they had been in in the previous study. Those who had played the role of needing help and had not received it (i.e., the other subject passed them by) were more likely to be helpful in the second experiment. These findings should give pause, since it seems likely that the positive aftereffects in this study, measured weeks later, could in other experiments have been negative aftereffects.

The knowledge that experimental participation can affect behavior for some time after the experiment should be a sobering thought for all social scientists. Debriefing does not always remove the effects of the experiment completely and thus cannot be overrelied upon to eradicate negative effects. Therefore it cannot be used to justify every deceptive study.

Alternatives to Deception
ROLE-PLAYING

One major alternative to deception is role-playing (e.g., Brown 1965; Hendrick 1977; Kelman 1967). In role-playing, subjects assume some role and try to behave as they would in a real situation. Role-playing may be passive, where subjects read a description of the experiment and predict how they would respond, or active, where subjects act out the role, often in surroundings that resemble the traditional laboratory setup (Mixon 1972, 1974). There have been criticisms of role-playing as a research tool (Berkowitz 1976; Freedman 1969; Miller 1972), and also spirited defenses of the technique (Forward, Canter and Kirsch 1976; Hendrick 1977; Mixon 1972, 1974; Stang, forthcoming). Cooper (1976) discusses the advantages and disadvantages of both role-playing and deception, and Hamilton (1976) discusses the two as noncommensurate methods.

In some areas where deception has been of little value, role-playing has been clearly useful. For example, qualities of airplane pilots and factors of cockpit design affecting human performance have been tested in simulators. The subjects, who may or may not be real pilots, act as if they are flying an airplane. It is certainly not necessary to deceive them into believing they are really doing so. Role-playing has been used to test theoretical predictions and practical human factors in a variety of areas, notably the Stanford prison study (Zimbardo et al. 1973), internation simulation games (Guetzkow et al. 1963), and exercises used in encounter groups and group dynamics settings (e.g., Moreno 1953). Role-playing has shown promise in a wide variety of studies (Mixon 1974; Stang, unpublished) and is certainly a major social science technique in its own right. Whether its findings can be generalized to everyday behavior depends on the situation, although in virtually every human situation people are acting out roles. When used to predict everyday behavior, role-playing may be more accurate if: the respondents' behavior is habitual, the situation is realistically recreated, the subjects' responses are similar to those to be predicted, and the demands and rules in the set are not drastically different from those in the everyday surroundings. Limitations will certainly be found in predicting everyday behavior from role-playing, but it should be noted that deception does not have a proved track record in predicting to everyday life either. In addition, theoretical development, not prediction to other behavior settings, is frequently the primary purpose of experimental studies. Role-playing thus can certainly be an efficient aid to theory construction and hypothesis generation (Cooper 1976) and may replace deception in some cases.

Role-playing has generally been criticized because the results of deception studies cannot *always* be replicated with this method (Darroch and Steiner 1970; Goldstein, Davis, and Herman 1975; Greenberg 1967; Holmes and Bennett 1974; Horowitz and Rothschild 1970; Nickel 1974; Willis and Willis 1970). The differences between role-playing and deception results are problematical only if one assumes that deception produces *the* answer. The methodological problems with deception experiments, as well as the lack of evidence that laboratory data can be generalized to everyday behavior, suggest that deception results are not the best criteria for decisions on the value of role-playing. Indeed, when Simons and Piliavin (1972) found that findings from role-playing subjects did not replicate deception subject results, they raised the question of which group's results were more ecologically valid.

Much more information is needed to assess the relative value of experimentation by means of deception and role-playing. We also need to know more about the roles subjects play in all kinds of experiments. Particularly in sophisticated forms, the research technique of role-playing has been underused and should be further developed. But just as certainly, the technique cannot be used for every scientific question. Researchers should be aware of the advantages and disadvantages of role-playing and must make an individual decision about its applicability in each instance.

FOREWARNING

Another alternative to total deception is to forewarn subjects that deception may be involved in the study. A researcher can ask subjects in a pool ahead of time if they are willing to participate in deception research and use only those who respond affirmatively (Campbell 1969*a*). When this precaution is not taken, the investigator may often forewarn subjects at the beginning of the session without compromising the study. This limited informed consent concerning the deception is ethically preferable to no consent at all, since the subject has agreed to be deceived. This removes part of the ethical question of direct lying, and research suggests that such a warning does not interfere with the accuracy of the data (Horowitz and Rothschild 1970; Holmes and Bennett 1974). Many subjects know that dissimulation may occur anyway, and stating this as a possibility at the outset of the experiment places the researcher in a much more ethically defensible position. Scientists such as Aronson and Carlsmith (1968) admit that most subjects know of social science deceptions; thus forewarning is a sensible strategy that can bring all subjects to the same level of sophistication regarding possible deceptions. Forewarning probably also allays retaliation effects by subjects who may be resentful over being surreptitiously deceived.

Conclusion

Deception studies have many potential ethical and methodological problems that must be faced squarely if one is planning to use the technique. One should carefully consider the following factors before conducting deceptive research:

1. Researchers should decide whether research deception is ever ethically defensible and whether a particular deception could have detrimental effects on themselves or on subjects. The potential results of the study must seem important enough to justify the ethical cost of lying.
2. The research should not create negative effects in subjects outside the realm of the study itself, such as cynicism or resentment or lessened altruism in real-life situations.
3. Deceptions should not be practiced when they are not necessary for effective completion of the research.
4. When subjects are exposed to potential harm or surrender substantive rights, informed consent is a necessity. Deception should not undercut subjects' rights to be informed beforehand about risks.
5. Safeguards such as debriefing should be used to minimize potential negative outcomes.

For further reading on particular questions see the references listed in specific sections of this chapter. Several references that are somewhat broader in coverage are: Aronson and Carlsmith 1968; Baumrind 1976; Berkowitz 1976; Kelman 1967; Mahoney 1978; Mills 1976; Mixon 1974; West and Gunn 1978.

II Ethics Related to Selected Nonlaboratory Approaches

6 Cross-Cultural Research

> Simply by being in a society he or she is studying, the anthropologist affects that society. By importing alien customs and novel equipment, he affects it even more. By importing opinions and values which are inevitably conveyed, he increases the effect. And, by being in a position to communicate findings and responses directly or indirectly to persons of authority or influence, the anthropologist may indeed drastically affect those he studies.
>
> Gerald Berreman (1973)

> The days of naive anthropology are over. It is no longer adequate to collect information about little known and powerless people; one needs to know also the uses to which that knowledge can be put.
>
> Wolf and Jorgenson (1970)

This chapter presents a general discussion of the ethical problems of cross-cultural field research, a technique frequently employed by anthropologists and other social scientists. The next chapter discusses the specific ethical questions related to one research technique, disguised participant observation. Researchers using the ethnographic method described here do not simply observe people in a single setting such as a laboratory, but study and describe life patterns using on-the-scene observation. Anthropological and sociological fieldwork usually consists of examining in detail the day-to-day life of some group. During the research the scientist often lives among those studied and participates in their daily activities. The research itself is usually a blend of intensive interviews, long-term observation, collection of artifacts, and questioning of informants. Anthropologists usually employ this approach to study groups in other countries, whereas sociologists use it to study groups within their own country. Psychologists and others occasionally use the ethnographic method for both types of groups. (See McCall and

Simmons 1969 for a methodological discussion of participant observation.) Although ethnographic research may focus upon the scientist's own group, usually the researcher examines other cultures or subcultures. However, the ethical principles that emerge will also apply to research within one's own culture.

Oscar Lewis's book *Five Families* (1959) is a good example of ethnographic research. Over a period of years, Lewis came to know the five Mexican families he described. In his words, "I have spent hundreds of hours with them in their homes, eaten with them, joined in their fiestas and dances, listened to their troubles, and discussed with them the history of their lives. They were generous with their time and good-naturedly submitted to Rorschach tests, Thematic Apperception Tests, the Semantic Differential, and intensive interviews" (p. 5). In other words, Lewis knew well the habits and feelings of those he studied. His book describes the activities of one day in the life of each of the families. Both the depth of ethnographic studies like Lewis's and the fact that they are conducted largely in other cultures create unique ethical problems. However, ethnographers still face the same problems as other researchers in reference to informed consent, privacy, and honesty.

Ethical Issues Shared with Other Methodologies

INFORMED CONSENT

As in laboratory research, the field observer is obligated to inform subjects of features of the study that might influence their desire to participate. It is especially important that the subjects be informed about what information the investigator will collect and how it will be used (Jorgenson 1971). There is a particular obligation for researchers to carefully explain the study and potential risks to uneducated subjects who may not initially understand how the research could affect their lives. If there is a danger that the country's government might use the anthropologist's report to control or exploit the group, this must be made clear (Tapp et al. 1974). In other words, the research and its implications must be clearly explained. The group and individuals should be allowed to participate freely and should not be coerced to cooperate by the government of that country or by the relatively large monetary inducements that scientists can sometimes offer in foreign countries.

William Helmreich (1973) presents a typical way that informed consent is acquired in field research. Over a period of six months he studied a black militant organization, but until he received permission from the head of the group, he kept no diary of his observations. The consent was given when the leaders of the group agreed to permit observations and the other members of the group knew of Helmreich's identity as a sociologist. Thus the group members could reveal or keep hidden whatever matters they chose. Obtaining consent is especially important when studies, like Helmreich's, explore private or sensitive areas.

PRIVACY

The field observer must respect people's right to privacy and recognize that definitions of privacy differ between cultures. He should therefore know what types of information and settings are most sensitive in the culture he will study. The group under study should be told whether individuals or the group as a whole will be identified in the published report or whether pseudonyms will be used. The participants should also be informed about the extent to which sensitive information may be published and whether this could have negative effects on them.

One traditional approach to protecting anonymity in field research is to use fictitious names for individuals, groups, or locations. In the past, pseudonyms for the groups and individuals studied were relied upon primarily for indigenous groups within one's own coutnry, whereas foreign peoples were usually explicitly identified as a group. At times pseudonyms have been enough to protect anonymity, but at other times they have not been successful. Recall, for example, the case of Springdale described in chapter 4. The town's residents were able to identify one another in the book *Small Town in Mass Society*, even though the sociologists had used pseudonyms.

Another case where pseudonyms were not enough to protect anonymity was Oscar Lewis's work, *The Children of Sanchez* (1963). Although Lewis had changed the names and locations to protect the family's anonymity (Beals 1969), the book caused such interest and controversy in Mexico that a big-city newspaper assigned reporters to track down the family. After a month's detective work they found them but fortunately did not disclose the name. Nevertheless this case, like the Springdale case, shows that pseudonyms may sometimes not insure absolute anonymity. The ethical researcher should examine the importance of anonymity in each study and the likely effectiveness of pseudonyms. Where group names and individual identities will be published, or where pseudonyms may not be sufficient safeguards, it is important to gain subjects' informed consent for publishing private material. However, even informed consent can be troublesome if subjects later change their minds (*Anthropology Newsletter* 1976).

There are many cases of published research in which the communities or individuals are not anonymous. Often it is necessary to identify the group so that other scientists may most fully use the information. In other cases it is impossible to protect the group's anonymity because too many others know where the scientist is working and hence will be able to identify the subject group later. For example, Rainwater and Pittman (1967) studied a large number of black families in a low-rent housing project in Saint Louis. The study was widely publicized in Saint Louis because the housing project had become a community scandal and many were seeking solutions to problems such as crime, accidents, and unemployment, which were rife within the complex. The publication of the findings could certainly have had repercussions for those living in the complex. The identity of the entire group could not be concealed, nor could the identity of the housing project officials described in the report. Hence the scientists had to face the possibility that their report might have serious consequences for those studied.

Rainwater and Pittman described activities of the residents of the housing project that the larger community would consider criminal, deviant, and immoral. Hence they faced the possibility that their report might serve to "strengthen prejudices and provide rationalizations for bigotry" (p. 361). They decided to present the findings openly, hoping that the truth would do more good than harm. Rainwater and Pittman partly solved the privacy problem inherent in their research by not promising subjects confidentiality, thus placing the decision to publicly reveal information on the respondents. But they realized that even this precaution did not completely release them from responsibility if the findings were misused, and so they sought to write their report in a way that was sensitive to the current political climate and to the needs of the community, including residents of the housing project.

Field researchers must be aware both of the subjects' definition of privacy and of how the published materials may affect the lives of those studied. Ethical precautions such as informed consent, confidentiality, anonymity, and the use of pseudonyms are often helpful, but even when these are employed, the investigator should evaluate the possible effects of research publication and be convinced that the subject population's rights and welfare are sufficiently protected.

DECEPTION

Three major types of deception have occurred in cross-cultural studies:

1. A social scientist or other person, usually in a foreign country, appears to be collecting scientific information but is really working for an intelligence operation such as the Central Intelligence Agency. Since Franz Boas (1919) castigated several anthropologists for working as foreign spies in World War I, social scientists have condemned using science as a cover for intelligence activities. A set of ethical guidelines for cross-cultural research prepared by Tapp et al. (1974, p. 249) enlarges upon the types of misrepresentation that are unethical: "Scientific research cannot rightfully be used to cover activities that have a different and hidden purpose, such as the gathering of military or political intelligence, of market information for commercial decisions, or of personal data for the evaluation of employees or other groups." Clearly, such deceitful operations would quickly make social scientists unwelcome abroad and in many domestic settings and would greatly curtail international social scientific inquiry (Beals 1967). However, as we shall see, social scientists may unwittingly act as agents for their governments. These cases present more difficult ethical issues. Professional intelligence agents may also pose as social scientists, and this cannot be controlled by scientists.

2. Another type of deceit occurs when the scientist disguises himself so that subjects do not know they are being observed. Usually the scientist takes on another role, and for this reason we call this type of undercover research *disguised participant observation*. Chapter 7 contains a detailed examination of the ethics of disguised research.

3. A third type of deception in field research is to mislead subjects about the aims or procedures of the research. One should strive to present the research truthfully, including facts about the sponsoring agency. As Tapp et al. (1974, p. 243) stated: "Openness and honesty, conducive to the creation and maintenance of trust, should

characterize the relationship between the investigator and the individuals and communities studied." They go on to state that mild deceptions in presenting research procedures are justified only after consultation with members of the host community and only if the subjects' welfare and dignity are assured.

It is often tempting for scientists to slant the description of their research to make it more appealing to the host community. For example, it is easy to emphasize to a government or tribe one's interest in an innocuous topic like language or kinship systems and to deemphasize controversial aspects such as the study of land tenure or social control (Barnes 1963). Although a small amount of "salesmanship" may not be unethical, complete honesty about the research aims and modesty about the potential benefits are usually desirable. It will certainly benefit social scientific research if investigators maintain a reputation as trustworthy and honest.

Ethical Problems Specific to Ethnographic Research
USES OF THE KNOWLEDGE

The cross-cultural researcher should be especially sensitive to the use of the published information. Scientific reports on other cultures and subcultures are often of interest to those outside the scientific community: to the military, government officials, employers, and competing groups. Often, the more information is known about a people, the more they may be manipulated. For example, in the study of residents of the Saint Louis housing project described earlier, Rainwater and Pittman realized that the information they collected would be of great interest to a wide range of persons who would use the knowledge for goals unrelated to science. They were concerned about presenting the information so that the entire community would benefit.

Anthropologists working in foreign countries have frequently collected or planned to collect scientific information that might also aid officials of that country's government or even foreign powers. The best-known case of research in which the information collected could have been used for military and political purposes by a foreign power was a project named "Camelot." In the mid-1960s, this six-million-dollar interdisciplinary project was funded by the United States Army in a contract with an American university (Silvert 1965; Vallance 1972). The university's research office arranged for leading United States social science scholars, including sociologists, political scientists, and anthropologists, to carry out the study, which was to be a large-scale investigation in Latin America on the causes of insurrection and the effects of various government actions in preventing revolt and revolution. The goal was to produce a theory and data that would help predict political and social change in underdeveloped nations (Horowitz 1967). The scientists were to analyze data from past revolts as well as contemporary societies. Countries like Mexico, Guatemala, and Argentina were to be carefully studied to determine the causes of insurgency, internal resistance, and rebellion against existing governments, and to determine how effective various governmental measures (e.g., violent suppression or small reforms) would be in quieting dissidents.

Although the United States Army obviously had a vested military interest in knowledge that would help forecast internal wars abroad, many leading social scientists were nevertheless attracted to the project. This interest developed because the huge scope of the project promised large returns in scientific knowledge and because the project was to be theoretical research with no promises to deliver secret data to the army. Some of the scientists were drawn by the unprecedented opportunity to pursue truth; others were attracted by the possibility of improving human conditions in Latin America. The populations studied were to be consulted throughout the project so that their voluntary assistance was guaranteed. Horowitz traced other reasons that many social scientists were willing to proceed with the project despite army sponsorship: they believed they would bring an enlightening, educational influence to the army itself; they believed they would have great freedom to handle the project as they wished; and they believed they would have an opportunity to do large-scale research with obvious relevance to real-world problems.

Less than a year after it began, Project Camelot was brought to an abrupt halt by the secretary of defense. When news of the project arrived in Chile, a hue and cry arose in the press over United States "intervention" and "imperialism." In the midst of the swirling debate that followed, the United States ambassador to Chile requested that the project be halted immediately, and demands at home for a congressional hearing soon followed. In this heated atmosphere the Department of Defense terminated the project, apparently because of pressures from the State Department and from President Johnson not to jeopardize United States foreign relations with such a potentially explosive research topic. And so

107

ended the largest social science project ever, funded by the United States Army and designed to investigate a politically sensitive area within foreign countries.

Although the scientists involved with the project had good intentions, they apparently did not consider the uses to which the United States Army or the Central Intelligence Agency could have put the information, such as fostering revolutions against regimes hostile to the United States. They failed to recognize the grave concern those in other countries would have over such potential uses. Camelot provides a prime example of the principle that scientists should, for reasons of both ethics and expediency, examine the probable extra-scientific uses of their data.

Other anthropological and sociological projects have also raised issues about sponsorship and possible misuse of data. For example, the United States Department of Defense sponsored a project on Himalayan cultures on India's borders. When Indian officials discovered that the project was funded by the United States military, they terminated their cooperation and enacted more restrictive policies for future research by foreigners within India (Berreman 1969). Another informative example was the United States support of social scientists who were studying the hill villages of Thailand. These villages were in strategic military positions for guerrilla bases and also could be influenced by the communists from neighboring countries. Therefore the Thai government and its ally, the United States government, had sound military reasons for wanting to understand the hill people. Although some of the anthropological research was directly funded by the United States Department of Defense (Wolf and Jorgenson 1970), many of the studies were done independent of military funding or objectives. However, as one of the researchers noted (Jones 1971), it made little difference who provided the funds—the United States and Thai governments used the findings anyway. As Jones said in regard to anthropological publications describing the hill people: "The more information there is available, the easier it is to develop new techniques of dealing with the people whom the government is trying to manipulate" (p. 348).

Regardless of the source of funding, all social scientists conducting research should be acutely aware of how their findings might be used in the current political and social climate. Funding should

108

certainly be refused if the researcher knows the sponsor will use the data to manipulate those under study. Funding should normally also be refused if a political, military, or commercial concern will keep the information secret so that it is inaccessible to other scientists. Tapp et al. (1974, p. 248) suggest that "A cross-cultural study should be undertaken only with the expectation that it will eventuate in full publication. If publication may jeopardize or damage the population studied, the investigator should delay publication or disguise identifying information, refraining from publication altogether only as a last resort." Jorgenson (1971) concluded that an evaluation of the way findings will be used may mean that sometimes the research should not be conducted in the first place because the information uncovered could bring physical or political harm to the people studied. The minimum safeguard in studies that may harm the group under study is that the subject population be informed of the risks and freely agree to accept them before the research begins. But this precaution still does not absolve the researcher from further responsibility.

In the past many scientists were naive about the possible detrimental influences their published findings might have for the people studied. Scientists must be more careful about interpreting data within the appropriate historical and cultural context in order to forestall misinterpretations. However, even a balanced presentation of findings is often not a sufficient safeguard. Much field research is done on minority and tribal groups, and the findings may be used by dominant groups for propaganda or to manipulate the community under study. For this reason, research publication has sometimes benefited those in power but harmed the minority peoples under study. Scientists need to consider this possibility because of the moral issue involved, but also because minority groups are themselves becoming aware that research may not benefit them or may even harm them. Understandably, then, they are often reluctant to participate and sometimes completely refuse cooperation. Failure to consider how one's findings may influence the people studied, given the current political situation, may thus harm scientific progress by creating a block to further studies on minority groups. Scientists should also recognize that their work might enhance the power position of some individuals or groups and should squarely face the ethical implications of this possibility.

109

ALTERING THE CULTURE

The ethical researcher should have an adequate knowledge of the culture, social structure, and customs of a community before the research begins. He should be sensitive and respectful toward laws, cultural expectations, and social customs while living in the community. In this way he will not blatantly violate local customs and will allow the everyday life of the community to proceed largely as it had before his arrival.

Traditionally social scientists have been concerned that they in no way alter the group they observed. Anthropologists and sociologists usually have tried to introduce a minimum of foreign influence by their presence. Social scientists have been reluctant to rush in and change the people they were studying for several reasons. First of all, their major work is science, not social work or missionary conversion. Second, the scientist is humble about his ability to "help" those in another culture, for he knows that a balance has usually evolved between various structures in a culture and that small attempts to "help" the groups might have unintended and unforeseen consequences, sometimes very negative, that would reverberate throughout the culture. There is humility in realizing that our own culture and values have shortcomings and may not invariably be beneficial to others. Last, the scientist has realized that his attempts at intervention, especially when not asked for, could represent an unwarranted imposition of foreign values.

The Tassaday, a simple and isolated people living in a remote region of the Philippines (Nance 1975), illustrate the difficulties faced by a researcher who sets out to help those in another culture. The recently discovered Tassaday had led an isolated stone-age, cave-dwelling existence for hundreds of years. For example, when they first saw a helicopter they were afraid it was a giant insect. They are a preagricultural gathering group who subsist on a diet of roots, fruits, tadpoles, and frogs. This unique assembly of only twenty-five persons is a laughing, loving group who are completely unaggressive and show a tender love for one another. The Tassaday are now protected by a government preserve and access to them by outsiders is limited. A scientist who decided to bring "progress" to the Tassaday would be presumptuous. Can the researcher be sure his values and goals would really help the Tassaday, or might not they utterly destroy their culture and all its positive attributes? Note that even relatively minor attempts to help the Tassaday might have major unforeseen consequences. Social scientists have usually attempted to observe groups like the Tassaday with as little interference in their lives as possible.

As the quote by Berreman at the beginning of this chapter suggests, we now realize that it is inevitable that the foreign observer will change the host community to some extent. The mere presence of an "expert" observer will influence how people act. This is true even when the observer comes from the same culture as the group being studied, and the influence can be stronger yet when a different culture is studied. In the case of the foreign observer, new values and customs will become familiar to the community and have an influence on it. Even such a seemingly minor change as giving the people steel axes may completely change the culture, including the social structure (Sharp 1952). In addition, the published results of the study are likely to have an impact on the group. Government administrators may read the materials when they formulate public policy. And it is becoming increasingly likely that persons within the community will read or learn what has been said in the report.

It may be unrealistic for the researcher to hope to exert no influence whatsoever, although he may still seek to minimize it. Perhaps a more realistic goal is to avoid harmful influences as far as possible. Realizing the limited state of our knowledge and the unpredictable effects a change may have on a complex and integrated cultural system, most scientists will be cautious about introducing changes into a foreign culture. Obviously interventions into a culture will be most ethical if they are requested by the host community and have no foreseeable negative effects. It is becoming increasingly clear that a scientist need not always refrain from action on behalf of those studied. Indeed, it is sometimes irresponsible not to intervene to help those in the host community.

INTERVENTION AND ADVOCACY

Scientists are obligated to help those they are studying when the host community faces an emergency or crisis. In these situations the scientist is morally obliged to intervene because the right to life, for example, is a higher value than the search for knowledge. For instance, imagine a psychologist studying child-rearing patterns who sees that a baby girl is malnourished and battered. According to an extreme value-free notion, the scientist might carefully record why and when the baby was beaten and what she was fed. But like the entymologist studying an ant colony, the scientist would not impose his values on the objects of his study by intervening in their situation. Such a scientist is obviously inhumane, and few scientists nowadays support such extremes in nonintervention. In fact, laws now generally require professionals to report such situations that come to their attention. The scientist in this situation has a conflict of values—those of acquiring knowledge versus the very life and welfare of an individual. Where the scientist has the ability to prevent some serious misfortune, his responsibility to do so takes precedence over the objective search for knowledge. For a situation in which anthropologists could choose to study a dying tribe or to attempt to gain aid that would save the band, Grayson (1969) says that "it is far more important to save life than record its passing" and that "the objective study of anthropology has no right to exist when it denies the right of existence to the living."

Another example of emergency intervention was Spillius's (1957) attempt to help the Tikopia during a famine. Spillius was working among the Tikopia (Polynesian Islanders), who were dying of famine because a hurricane had destroyed their crops. Stealing became common and a political crisis ensued. Open violence was averted because Spillius worked to help local leaders solve their problems. All sides recognized the need for help and were seeking solutions to the famine; therefore he was not imposing outside values and goals on the group. Spillius argued that the researcher must assume professional responsibility for his actions and leave the group better able to cope with their problems after the scientist has gone. Helping people must be done carefully and in culturally acceptable ways so that the researcher does not do more harm than good by his intervention. At times the field researcher will intervene to help those he is studying because they request aid and are unlikely to solve the problem without his intervention. Gallin (1959) found it impossible to remain a neutral observer. He was studying a small village in Taiwan when two village farmers were badly beaten up by nearby villagers. The townspeople were generally unable to achieve justice because of their disorganization and ignorance of legal procedures, and they requested Gallin's aid. Because of considerations of justice and to keep good rapport with the villagers, Gallin did help the victims in the legal process, although he tried to intervene as little as possible. Gallin found that this intervention greatly aided his anthropological work because the villagers became more open in their revelations to him and more willing to answer his questions. He himself raised the ethical question of whether he could take knowledge from the village and give nothing in return when requested to do so.

There is emerging in the specialty of "applied anthropology" a model of helping those studied while simultaneously doing research on cultural change. Those conducting research of this type—also called "action anthropology," "participant intervention," and "operational research"—seek to combine theoretical and practical goals. For example, in 1948 Gearing (1960) and his colleagues started the "Fox Project," aimed at helping the Fox Indians build better relations with their Anglo neighbors. Probably the largest applied program by anthropologists was Project Vicos. In 1952, social scientists from Cornell University took over Vicos, a thirty-thousand acre hacienda in Peru, for five years. Holmberg (1958), an anthropologist, assumed responsibility for and authority over the two thousand Peruvians on the hacienda. The project was sponsored by the Peruvian government to transfer ownership and control of the land from a single autocratic *patrón* to the Indians who worked the land. During their term as *patróns*, the anthropologists helped institute a democratic government to manage community affairs. In addition, they initiated a program of education and training for the Vicosinos. Modern methods of agriculture were introduced, and food production was increased 500 percent in five years. A health clinic was built by the Indians and a public health program was inaugurated. Obviously the project made a phenomenal change in the host community.

Holmberg (1958) has argued convincingly for the justifiability of applied anthropological work. He maintains that the goals of the Vicos project were self-determination and a wider sharing of the resources and power within the community. Other major aims were to increase the self-respect and dignity of the Vicosinos and improve their health. Holmberg argues not only for the humanitarian benefits of such applied projects, but for their scientific merit as well. He said, "I remain convinced that the interventionist or action approach to the dynamics of culture, applied with proper restraint, may in the long run provide considerable payoff in terms both of more rational policy and better science" (p. 12). The scientist who contemplates initiating changes in another culture should realize that the entire impact of the intervention is not foreseeable and that

113

the host community may have a different value perspective. Normally, attempts to help a group should follow a request from the group itself and a careful analysis of the situation.

BENEFIT TO THOSE STUDIED

As is evident from the above examples of interventionist research, cross-cultural researchers are cautiously moving away from the neutral observer model and often want to help those they study. There is a growing belief that the host community should benefit from research because they have contributed to the project and because benefit to subjects will encourage closer cooperation between scientists and those they study. Minority peoples and tribal groups are becoming increasingly annoyed at being simply objects under the scientists' microscopes. Communities are mainly concerned not with the advancement of an academic discipline but with their own welfare and development. Subject groups, who are often disadvantaged, increasingly want to know how the research will benefit them. For example, many black communities want to know how proposed research about them may help break down institutional barriers to their advancement. Similarly, native American tribes have asked how the anthropologist can help them.

Vine Deloria, in *Custer Died for Your Sins* (1969), fires blistering criticisms at anthropologists for studying Indians for years without helping them develop. Even though Deloria's criticisms are somewhat unfair to anthropologists who have helped native American tribes, his critique clearly expresses the heightened belief among minorities that research should be not just *on* them, but also *for* them. As Deloria wrote, "We should not be objects of observation for those who do nothing to help us. . . . Why should we continue to provide private zoos for anthropologists? Why should tribes have to compete with scholars for funds, when their scholarly productions are so useless and irrelevant to life?"

The researcher may help the group he studies in a variety of ways. Organized knowledge about the group may be of interest and help in itself. An example is the study on the Ashanti by Rattray. Rattray (1956) wrote down the customs of the tribe, and his work thereafter served as the basis for law in that culture. The scientist may also aid the community by technical assistance, advocacy on its behalf, consulting experts about local problems, and so forth. One way scientists may make it more likely that the research itself will benefit the group is to plan the entire project or parts of it with representatives of the host community. In the words of Tapp et al. (1974, p. 245):

> The total research experience should enrich and benefit in some way the individual subjects and host community. A major source of potential benefit is generating knowledge that may enhance the welfare of the population studied and the development of the community. To maximize these potential benefits, members of the host community should be involved in all phases of the research. Consider, in setting research priorities, not only theoretical problems of the discipline but also needs of the host community for research related to its welfare and development.

Conclusion: From Academic Colonialism to Reciprocity and Respect

Despite the wisdom of caution when intervening in other cultures, often the scientist will feel a moral obligation to help the group. As we stressed earlier, the scientific endeavor should have mutual benefits for the scientist and the host community. The scientist incurs an obligation to try to meet some of the group's perceived needs. He should consider the welfare and development of the group and should discuss the project plans with members of the host community. The scientist should not rush in and "straighten out the natives," but neither should he totally ignore the needs of those under study, especially since host communities are increasingly requesting that research have some direct payoffs for them.

For additional discussions of ethical issues in cross-cultural research, see: American Anthropological Association Code of Ethics; Adams 1971; Barnes 1973; Beals 1969; Rynkiewich and Spradley 1976. Sjoberg 1971; Tapp et al. 1974; Vargus 1971; and Weaver 1973.

7 Disguised Participant Observation

> What I propose, then, at least as a beginning, is the following: first, that it is unethical for a sociologist to deliberately misrepresent his identity for the purpose of entering a private domain to which he is not otherwise eligible; and second, that it is unethical for a sociologist to deliberately misrepresent the character of the research in which he is engaged.
>
> Kai Erikson (1967)

> Most of us, in fact, never cease observing the social sphere about us and are continually interpreting the behavior of people about us.
>
> Julius Roth (1962)

One of the most exciting research techniques used in natural settings is in-role participation by the social scientist, a "method in which the observer participates in the daily life of the people under study, either openly in the role of researcher or covertly in some disguised role" (Becker and Geer 1957, p. 28). Kluckholn (1940, p. 331) has defined participant observation as the "conscious and systematic sharing, insofar as circumstances permit, in the life-activities, and, on occasion, in the interests and affects of a group of persons." Following chapter 6, which covers ethnographic types of research in general, this chapter focuses more narrowly on participant observation in which the researcher disguises his true identity and conceals the research endeavor.

When social scientists participate in the daily lives of those they are studying and their roles as researchers are known to those they are observing, they are studying behavior through participant observation. Those who utilize the technique have two major goals: to gain a new perspective on the lives of those observed by personally participating with them, and to make direct observations of the group's behavior instead of relying on verbal reports. In addition, a participant observer is privy to gossip and other informal conversations that reveal much about the unstated goals and attitudes of the group. In contrast to open participant role-taking, the scientist who disguises his occupation and research activities becomes involved in *disguised observation*. Erikson's statement at the beginning of the chapter is an example of the strong ethical objections some scientists have toward disguised observation, even to the point of recommending the total abolition of the technique. We will see that disguised observation is plagued with some of the same ethical shortcomings as experimental deception, for example, lying, lack of informed consent, and the invasion of privacy.

The Disguised-Undisguised Continuum

The disguised and undisguised forms of participant observation are not completely distinct, but can be placed along a continuum of openness, as the following examples will illustrate. One case of open, nondeceitful participation involved George Kirkham (1975), a social scientist who left his university post as professor of criminology and joined the Jacksonville, Florida, police force to "develop greater insight into the police role in modern society" (p. 18). Kirkham was most interested in his own experiences and feelings as an officer, not in spying on others or infiltrating the private domain of police life. In fact the other policemen knew Kirkham's background and called him "Doc." When he joined the force, he planned to show how an educated and understanding police officer could handle situations in a more enlightened way than the proverbial "pig." Having lived through the student-police clashes at Berkeley and being concerned with police mistreatment of minorities, Kirkham was determined to demonstrate how a model officer should act. In order to provide himself with the greatest challenge, he requested assignment to a tough ghetto area with a high incidence of violent crime. Within five months on "beat 305," Kirkham changed dramatically. In his words, he became filled with "punitiveness, pervasive cynicism and mistrust of others, chronic irritability and free-floating hostility, racism, a diffuse personal anxiety over the menace of crime and criminals that seemed at times to border on the obsessive" (p. 19). The longer he served in the high-crime area, the more conservative his attitudes became and, unfortunately, the more hardened he felt toward blacks. When Kirkham later began working in a low-crime suburb, he became a benevolent "good-guy cop," the type of officer he previously thought

117

all policemen should be. Kirkham's personal experience demonstrates the strong situational forces that can shape the attitudes and behaviors of law enforcement personnel, and it gave him a much deeper understanding of police life. The study was conducted without deceit or disguise. Although he was also a social scientist, Kirkham was a real policeman, having graduated from the police academy and served as a regular uniformed officer on patrol.

The amount of concealment in field observation varies along a continuum from Kirkham's openness to complete impersonation. It is difficult to say just where honesty ends and disguised observation begins, as the following example illustrates. In a famous book, *Black Like Me*, John Griffin recounted his experiences as a black man (1961). Griffin, a white man, was sickened by the treatment blacks received in the United States and decided to experience firsthand what it is like to be black. He intended to gain a unique perspective on a problem no white had and to write more authentically about the problems faced by blacks. Griffin shaved off his body hair, took drugs that darkened his skin, and aided the pigmentation process with dye. His book recounts his experiences while living as a "black" in the South for several months. The inhuman treatment Griffin frequently received shocked Americans who read of his experiences. Although he was not a social scientist by profession, Griffin's case points up the lack of a clear line between open observation and disguised techniques. Griffin colored his skin and let people assume what they would. He did not lie, but his appearance was obviously calculated to mislead others about his race.

This study also demonstrates the uncertainty about when a disguise damages other people's rights and becomes unethical. Griffin merely changed his appearance and observed others. In fact, scientists may routinely make observations about people without informing them. No disguises are involved; it is just that social scientists constantly observe others' behavior. These everyday observations of our family, friends, and self frequently become parts of the theories or data of the researcher. Certainly the social scientist cannot wear a warning sign: "You may be the subject of scientific observation." We can see, therefore, that many of our observations are not "open" and known to those we are observing. The ethical issues of privacy and informed consent occur even for undisguised participation. But when disguised participant observation is employed, additional ethical problems arise. Ethical questions become more important as we move toward the disguised end of the continuum. However, because it is a continuum it is impossible to draw absolute ethical guidelines specifying a particular point where participant observation studies become unethical.

Ethical Considerations of Undisguised Participant Observation

One ethical issue involved with undisguised participant observation is that as people become comfortable with the researcher, they may forget that he is recording their behavior. Although this is one of the major advantages of the technique, it means that people may reveal opinions or information they do not want recorded. This problem is most acute when the group is talking casually, for example, over a drink after work, because subjects are likely to think of the researcher as relaxed and temporarily out of the research role (Davis 1961). Few people realize that social scientists almost never stop observing and analyzing behavior, no matter how naturally they fit into a group. Only the most scrupulous investigators will ethically object to such invasions of privacy, but it is often desirable to make the research report or selected portions of it available to the subjects at the conclusion of the study, giving them the opportunity to object to any reported material they may have given "in confidence." The report should be written to protect the anonymity of the group and the individuals, unless the participants agree to waive this before the study begins. Usually a combination of full disclosure of one's role as a scientist, making the preliminary report available for participants' objections to confidential material, and protecting the anonymity of the group will eliminate any ethical problems. Whenever confidential material will be published, or when participants' anonymity will not be protected, prior informed consent should usually be obtained.

Ethical considerations aside, overt participant observation has several practical and methodological drawbacks, which make disguised observation more useful in many cases. One such practical disadvantage concerns the possibility that subjects may not admit scientists to their circle if they know their true intent and identity. In addition, if the subjects know the investigator is a scientist, they may purposely alter their behavior or conceal facts that are known to group members. These shortcomings can be overcome by disguised participation. However, as will be seen, even in disguised participant observation the scientist may influence the behavior of the group. Another methodological problem common to both undisguised and disguised participant observation is the possibility that the observer may lose his objectivity. For example, two sociologist observers at a Billy Graham crusade were so swayed by the evangelist's oratory that they were converted and marched to the front to make a profession of faith (Webb et al. 1966). Certainly their status as objective observers became jeopardized at that point!

Ethical Considerations of Disguised Participant Observation
Disguised observation techniques seem to share some of the excitement of methods used by detectives, investigative reporters, and CIA agents. They are designed to provide access to groups even when the groups do not want a scientist in their midst (e.g., the Mafia). Since subjects will often continue to behave naturally, the investigator will be given the opportunity to obtain relatively non-reactive observations. Despite the advantages of providing un-obtrusive observations and access to otherwise unobservable spheres of action, disguised participation still shares some of the problems of more open participation. Scientists may become involved and emotionally caught up in the activities of the group and lose their objectivity, or they may influence the behavior of the group they are observing, as later cases will demonstrate. Aside from methodological problems, those using disguised participation must face the difficult ethical problems described below.

DECEPTION
The ethical issues related to experiments using deception clearly apply to the disguised participation technique, as the following example demonstrates. In a study by Rosenhan (1973), a number of observers recorded their observations of life in mental hospitals by becoming psychiatric patients themselves. Rosenhan sent eight "pseudopatients" into mental hospitals all over the United States to claim that they were hearing voices that said "hollow" and "thud." The pseudopatients were in fact mostly professionals, and none of them had serious psychological problems. All were immediately admitted to psychiatric wards, where they openly took notes on hospital activities. When the staff asked how they felt, the pseudo-patients responded that they were fine and that the symptoms were gone. While real patients often believed the pseudopatients were

journalists or professors because of their constant note-taking, the staff interpreted this "writing behavior" as another manifestation of psychopathology.

The pseudopatients, with only one exception, were diagnosed as schizophrenic and kept in the hospitals from seven to fifty-two days. When released, they were considered "in remission." The study demonstrated hospital staffs' inability to detect "sane" people as well as their bias toward interpreting patients' behavior as pathological. But though the study provided an important insight into mental hospitals, it did so at the cost of deceiving a large number of staff. Debriefing to help the staff members learn from and accept the study would possibly have made it more ethical. Other considerations applying to deception, such as the magnitude of the lies, presence of consent from the individual or group, whether subjects' privacy or other rights are jeopardized, and whether safeguards are used to protect subjects' well-being should also be employed to assess whether a disguised role study can be ethically justified.

PRIVACY

Although disguised role studies and deception studies obviously share many of the same ethical problems, disguised participation is not included in the chapter on deception because this method raises unique ethical concerns, particularly with respect to privacy. The disguised participant, for example, may be admitted to a private setting and learn of private information that subjects would not reveal if they knew he was a scientist. The disguised participant may even impersonate a friendly comrade and use interpersonal trust and group spirit to acquire information from unsuspecting subjects, as is illustrated in a study by Sullivan, Queen, and Patrick (1958) on military training. The United States Air Force, wanting to get a better idea of how new airmen experienced induction and training, decided to disguise an Air Force officer as an enlistee and have him go through recruiting, induction, basic training, and technical training. The officer was given a new identity and a fictitious personal history and was trained for nine months in his "new personality." He lost thirty-five pounds, underwent minor surgery to make him look younger, and was provided with a "coach" to train him in the speech patterns and dress style of the "adolescent subculture." The officer's changed appearance was so convincing that the sergeant at the recruiting station (ignorant of the study) recommended that the boy not be accepted by the Air Force because he appeared to be a juvenile delinquent. On the first day the observer defied an instructor's order. When other trainees disobeyed rules and covered up jobs they had not done, the officer-researcher did so too. Since "Tom" gained the friendship and trust of his fellow soldiers, he was able to uncover private information about their attitudes and about "tricks" they used. The researchers justified this invasion of privacy because they firmly believed "Tom's" complete integration in the trainee group was necessary for the project's success.

Although few disguised participant studies rely on such extensive deception, most do invade private spheres to some extent, and such invasions require careful prior analysis. Will safeguards be taken to protect subjects' anonymity? Will the invasion of privacy cause protests among the public when the study is published? These are several of the questions an investigator must honestly face when using this research method. In defense of the Air Force study, it should be said that the authors reported that the data were safeguarded so that no disciplinary action could be taken against those in the study. In addition, as noted in chapter 1, persons in a military setting have already relinquished many rights, including privacy, possessed by ordinary citizens. See also Coser (1959) and Queen (1959) for a debate on the ethics of this study.

INFORMED CONSENT

The third major concern with disguised participant observation is that the subjects have no opportunity to exercise informed consent. For instance, Laud Humphreys, a minister turned student sociologist, studied homosexual behavior in public restrooms, participating himself as a "voyeur" whose job was to warn the men if someone was approaching. Humphreys (1970) carefully observed and recorded a large number of cases of fellatio and a lesser number of other homosexual behaviors. He observed the signaling procedures used by homosexuals interested in one another, the details of the sexual interaction, and demographic characteristics of the males involved. The activity was extremely sensitive because it was illegal and none of those involved would have wanted the information made public. The homosexuals used isolated lavatories for their activities and immediately terminated the contact if someone else approached. Humphreys was allowed to observe the proceedings as a "lookout" because those recognized by the homosexuals as voyeurs were allowed to watch. The researcher did not take a more active role as participant observer but faithfully performed his duties as lookout.

Humphreys also recorded the automobile license numbers of men who frequently came by car and gained their names and addresses by presenting himself at the Department of Motor Vehicles as a market researcher. After about a year, Humphreys interviewed the homosexuals in their own homes, changing his appearance and automobile so the men did not recognize him. He did not indicate that he knew the subjects were homosexuals but pretended they had been chosen randomly for a survey. As soon as possible after the study was over, Humphreys destroyed the names of his homosexual subjects so that their identities could not be traced. Laud Humphreys's research illustrates that even when one takes precautions to protect subjects, it is usually impossible to gain their prior informed consent in a disguised observation study. The lack of informed consent and other questionable procedures in his study infringed on subjects' rights.

INFLUENCING THE GROUPS OBSERVED

The participant observer must always face the methodological and ethical problem that his behavior may affect those studied. On the methodological side there is danger that the results will not be representative of natural events because the observer and his actions are a foreign influence that might alter the flow of events. Both Mead (1961) and Erikson (1967) have argued that disguised observers inevitably change what they are observing because they give subtle nonverbal cues of their deceit and because they cannot possibly react just like real participants. This possibility often also represents an ethical dilemma, since one is influencing the behavior of others, sometimes strongly, without their knowledge or consent. The problem is even more acute if the scientist is reinforcing deviant or antisocial behavior. The Festinger, Riecken, and Schachter (1956) study of a spaceship cult mentioned in chapter 5 illustrated the impact scientific observers can have on a group. As the authors said, "Our initial hope—to avoid *any* influence upon the movement—turned out to be somewhat unrealistic for reasons outside our control and inherent in the process of making such a study as this" (p. 237).

Festinger and his colleagues hired observers to infiltrate the somewhat secretive group, arming them with personal "psychic" stories to help them gain entry. One told of a "dream" that seemed to prophesy the great flood predicted by the group; another hired observer told the group of a fictitious supernatural experience he had in Mexico. Both stories heightened the faith of the cultists. When several disguised scientific observers joined the group in a short time, the group took this as a sign that the spacemen were supporting their movement. Every act of the observers within the group had some meaning—usually affirming the cultists' beliefs. The researchers stated, "In short, as members, the observers could not be neutral—any action had consequences" (p. 244). There are a number of conditions under which observers are likely to exert a considerable influence on the group—for example, when the group is small and when there is a great deal of interpersonal interaction.

ADDITIONAL PROBLEMS WITH DISGUISED OBSERVATION

Erikson (1967) has charged that a major problem with disguised participant observation studies is that in most cases student research assistants, not trained scientists, actually participate and observe. He noted that in a series of disguised observation studies he reviewed, almost all were carried out by graduate students. For example, in a study of Alcoholics Anonymous groups done by Lofland and Lejeune (1960), graduate students were sent to the meetings posing as alcoholics seeking help. Placing students in these roles presents several ethical problems. A student might feel forced to accept the role, afraid of being confronted by his supervisor, even though he felt uneasy about the study. This is a problem whenever a researcher directs others to carry out ethically questionable studies. Also, assistants are often in an embarrassing and sometimes vulnerable position if they are exposed. Furthermore, because of the lack of education about ethics and the pressure to collect data, breaches of anonymity or privacy may be more likely when apprentices conduct the actual research. This possibility points to the need for close supervision and training of research assistants used in such studies. Assistants should be given a choice whether they wish to take part in a study with any questionable ethical features, and the investigator should insure that they are aware of the issues involved before they decide.

124

As with all deceptive procedures, disguised observation may offend society because of the deceit and invasions of privacy, and such reactions could generalize to the entire social scientific enterprise. One critic of the Lofland and Lejeune study (Erikson 1967) suggested that angry reactions from the Alcoholics Anonymous organization could close the doors to future researchers who approached AA groups in a more open and honest way. So far, the reporting of disguised participation studies has not aroused public anger; thus this caveat may be crucial only for those who delve into very sensitive topics or groups.

Conclusion

Since disguised and nondisguised participation studies range along a continuum, there is no absolute cutoff where they definitely become unethical. Nor is there agreement among social scientists about whether disguised participant observation is justified. Erikson (1967), Jorgenson (1971), and Davis (1961) have argued that disguised participation is unethical, whereas Denzin (1968), Roth (1962), and Galliher (1973) defend the strategy. Galliher even argues that secret research against "bad" organizations such as the Ku Klux Klan and the Pentagon is a moral necessity. Because the ethical issues are manifold and complex and because the degree of openness may vary greatly, there can probably be no ethical consensus about disguised participation techniques in general. Each investigation should be examined for the amount of deception, extent of invasion of privacy, and other ethical problems, and ethical decisions should be made individually.

Ethical Suggestions

It is apparent that complete abandonment of all levels of disguised observation would be a large loss; but it is equally apparent that the method has many ethical shortcomings. If a researcher examines the ethical issues and decides to proceed with a study, several safeguards should help minimize detrimental effects:

1. Deceive as little as possible. Do not deceive subjects when it serves little purpose, and remember that it is better to let others draw incorrect conclusions or simply be ignorant of your scientific study than to actively lie.

2. Enter private spheres with the maximum informed consent

125

consonant with the research goals. When a private sphere has been observed without participants' knowledge, obtain their informed consent post hoc whenever possible.

3. Plan procedures that absolutely guarantee subject anonymity, especially in published reports. One safeguard, especially where sensitive information is reported, is to let the subjects read a draft of the report and point out any passages they feel were given in confidence or any statements that might jeopardize their anonymity.

4. Review the potential influences of the observers on the group and rework the study if any negative consequences are foreseen.

5. Fully inform research assistants about the research before it begins and give them free choice whether to participate.

6. Consider whether the study could cause indignant outrage against social science, thus hampering other research endeavors. If so, consider canceling the study or using a different methodology.

7. As with all procedures that have serious ethical questions, consult colleagues and request their suggestions for minimizing ethical problems. If possible, consult representatives from the group to be studied.

For additional reading see Denzin 1968; Erikson 1967; McCall and Simmons 1969; Shils 1959.

8 Experimental Interventions and Evaluation

> Therefore it is our judgment that rather than regarding institutional evaluative research as a potential violation of resident rights, these descriptive and comparative research efforts should be *mandatory* aspects of responsible institutional management.
>
> Davison and Stuart (1975)

> This chapter is committed to the experiment: as the only means for settling disputes regarding educational practice, as the only way of verifying educational improvements, and as the only way of establishing a cumulative tradition in which improvements can be introduced without the danger of a faddish discard of old wisdom in favor of inferior novelties.
>
> Campbell and Stanley (1967)

Experimental intervention is an activity planned and implemented by a scientist to alter the behavior, thoughts, or feelings of an individual or group. The scientist introduces an experimental "treatment" intended to produce positive changes in a target population and then measures the changes. This chapter considers such experiments, carried out in everyday life situations, and the studies evaluating their results. Such experimental interventions may be of many types; educational innovations, community change, individual therapy, and social policy alterations are examples of "real world" change attempts.

Numerous ethical issues are raised by experimental intervention attempts. Some of these, such as those concerned with informed consent and privacy, are common to most types of research. But additional ethical problems tend to be encountered when experimental changes are attempted. For example, there is the issue of whether the scientist has the right to impose his values and attitudes on those he is attempting to change. Another important question is when untreated control groups are ethically justifiable. The follow-

127

ing examples will illustrate some of the ethical questions faced by those who propose social science experiments that may directly alter people's lives.

Case Studies: Experimental Intervention

The history of medical experimentation has shown that ethical problems may occur in real world experiments (Beecher 1966*b*, 1970). For example, in a medical study on the treatment of syphilis a control group of infected men was left untreated. Even after the experiment was over and remedies for syphilis had been discovered, these men did not receive treatment. They were forgotten by the experimenters, and their health deteriorated for many years. In another experiment, during World War II, radioactive plutonium was injected directly into the bloodstream of eighteen men, women, and children without their knowledge. It was believed at the time that even small amounts of plutonium could cause cancer. In a third study, conducted by the CIA and the United States Army, subjects were given the hallucinogen LSD without their knowledge; at least one man subsequently became very disturbed and committed suicide. Experimental interventions within the social sciences and the resultant ethical problems have been much less extreme than these medical examples. But as the following cases illustrate, ethical issues do arise even in studies that do not appear to directly harm the participants.

EDUCATIONAL EXPERIMENTS

Many experiments are conducted in our schools each year—from preschools to graduate schools. Educational researchers introduce new social climates and methods of learning, and evaluate their effects. A study that demonstrates several of the ethical problems educational researchers confront was conducted by Fraser et al. (1977). Fraser believed that most college classes are too competitive and that learning would be enhanced if students were more cooperative and helped one another more. In "tutoring" one another, Fraser argued, the learner would receive an explanation of the material and the tutor would have an opportunity to review and organize the material. Another benefit of a cooperative classroom, he believed, would be that students would prod each other to study harder. Fraser reasoned that cooperative "peer monitoring" would be

128

helpful to learning and would be expedited in classrooms where grades were based on group, not individual, performance. This study was based on group cooperative learning methods employed in Russian classrooms (Bronfenbrenner 1973).

Fraser instituted a peer-monitoring experiment in several social psychology classes he was teaching. In one class students were matched in pairs and their course grades were dependent on how they performed together (their average performance). If one student "earned" an A and the other achieved a C, they would both be given Bs in the class. Students were matched according to their grade point averages (GPA), with one high- and one low-GPA student or two medium-GPA students in each pair. Students were told, however, that the matching was done randomly. Fraser hypothesized that both partners would profit, if not in grades, at least in the amount learned. A comparable class served as a control and was taught exactly the same except without peer monitoring. It was found that those in the peer-monitoring condition performed substantially better on exams than those in the control class, suggesting that grading contingencies fostering cooperation did enhance learning. Whereas 11 percent of those in the control class received a D or an F, no student in the experimental class did so. As it turned out, not a single student's grade was lowered by his partner.

In a second study, Fraser randomly assigned students from within the same classroom to grade-contingency groups of varying sizes. Students were informed about the peer-monitoring procedure during the first week of class and given an opportunity to drop the class. If they remained in the class, they were required to participate in the study and accept the partnership assignment given them. Some students were assigned to groups of two, others to groups of three, and still others to foursomes. The members of each group, regardless of group size, received the average grade earned by that group. Once again Fraser reasoned that good students would be motivated to help the poorer students within their group. A number of randomly selected students were not assigned a partner and thus served as a no-treatment control group. Those in peer-monitoring partnerships performed significantly better than did the controls. In this study six students (out of 179) received a grade one step lower than they would have under traditional grading.

These studies by Fraser et al. demonstrate a whole host of ethical

129

issues in experimental treatment programs. Can subjects be forced to participate in a new program without their free and informed consent? Was freedom sufficiently guaranteed by allowing students to drop out early in the quarter? Could an instructor justify using new grading practices for research purposes? For example, several other faculty members raised questions about the fairness of the grading because students might receive a grade they did not "deserve." What are the ethics concerning students who were harmed by the treatment (e.g., had their grades lowered)? Conversely, was it fair to students in the control groups who did not have the benefit of what seemed to be a valuable experimental treatment? Were control-group subjects unjustly deprived of a learning aid? Educational studies on new techniques are likely to raise some of the same ethical issues posed by Fraser's study on peer monitoring.

It may often be helpful to divide the treatment program from the evaluation component and to perform separate ethical analyses of the intervention and of the measurement of the results. For example, in Fraser's study few students objected to the experimental assignment of subjects or to the research evaluation of the procedures. The complaints voiced mainly concerned the treatment program (peer monitoring), and this could have occurred without any assessment research. Therefore most of the ethical questions concerned the innovative treatment, not the research itself. In other cases the measurement component may raise ethical issues whereas the intervention does not.

The Negative Income Tax

Policymakers are well aware of the profound changes their decisions can create. Only recently have social scientists begun to aid them by measuring the effects of new programs. In the last decade the United States government has funded a series of controlled experiments on new social programs, heeding Donald Campbell's (1969b) call for an experimenting society. The idea is that if new government programs are carried out on a restricted local scale, important factors can be manipulated and the results measured. Thus new programs can be modified or abandoned if the program proves unsuccessful. Previously, new social programs were often instituted nationwide, but their success was usually debatable in the absence of adequate measures. Furthermore, it was often difficult to abolish

130

these huge programs even when they appeared to be unsuccessful. It is clearly desirable to understand beforehand what changes a new program will bring about, whether the program involves new preschools, housing developments, or social welfare reforms.

The New Jersey negative income tax experiment was one of the first major government-funded social policy studies. The idea was that money would be paid (negative taxes) to low-income families. If the family's income rose above some minimal level, the negative taxes would be reduced as income climbed. Thus families would be guaranteed a base income, but because the payments were paid on a graduated percentage above this basic level, they could always receive more money by working. For example, a family would be guaranteed an income of $3,000 annually, but above this level payments might be reduced to 50 percent of earned income. This graduated payment feature was aimed at minimizing the lack of incentive to work that is often associated with welfare programs. One purpose of the New Jersey experiment was to assess whether the negative income tax would decrease the participants' desire to work.

Thirteen hundred low-income families in New Jersey and Pennsylvania were selected to participate in this ten-million-dollar experiment (Kershaw 1972). About eight hundred of these were in experimental groups, receiving varying amounts of income support. Five hundred families served as control participants, receiving no income payments. The participant families could spend the money however they wished. The participants intermittently completed questionnaires and underwent interviews, from which their work behavior was determined. The control subjects were paid a small amount for their participation. The results showed that compared with the control group, those receiving income support generally did not earn or work significantly less (Kershaw and Skidmore 1974).

Partly because of the success of the New Jersey experiment, a series of governmentally funded income-maintenance and family-assistance experiments are now being conducted around the United States. Other social program experiments are being carried out on health insurance, housing vouchers, and school vouchers. Ethical questions about informed consent, privacy, and untreated control groups are often associated with such social experiments (M. S. Frankel 1976). And there is another important question: Who should decide the goals underlying social policy experiments?

EVALUATION RESEARCH

Evaluation research is a broad class of methodologies and view-points designed to evaluate the adequacy of institutional programs. The programs may or may not be carried out under the direction of behavioral scientists and may or may not be designed as formal experiments. Evaluation researchers are usually employed to aid in the assessment of large-scale institutional programs. Many of the ethical issues discussed in this chapter apply to evaluation research, and questions of values and politics are extremely important. As Lee Cronbach (1977, p. 1) wrote, "If evaluation is not a scientific activity, what is it? It is first and foremost a political activity, a function performed within a social system." The broad area of evaluation is growing fast and generating many practical and ethical problems for researchers, policymakers, and society as a whole. At the same time the importance of this work cannot be overempha-sized. In addition to the ethical issues discussed in this chapter, further material on ethical and value issues in evaluation research may be found in *Evaluation*; Guttentag and Struening 1975; House 1973, 1976; Scriven 1976; Sjoberg 1975; Struening and Guttentag 1975; and Suchman 1967. See also the discussion of applied anthropology in chapter 6, which combines intervention with research.

EXPERIMENTS IN COMMUNITY CHANGE

Social scientists are becoming increasingly involved in attempts at community improvement (Riecken and Boruch 1974). Working within such settings as judicial systems, recreation programs, and community centers, scientists have sought to make changes and measure the effectiveness of their efforts. In the role of community change agent, the scientist is not merely a passive recorder of events; he works actively to achieve certain goals and values in the commu-nity. A project conducted by Seidman, Rappaport, and Davidson (1976) with juvenile offenders illustrates a community change program. Seidman and his colleagues wanted to divert teenage offenders from the typical juvenile court and probation procedure because they believed this system could make delinquency problems worse by labeling the child as a delinquent and putting him in contact with other youthful offenders. The researchers' goal was to employ undergraduate college students to work on the youths'

132

behalf. Although the investigators could not be certain such an intervention would be more effective than the court system in curtailing delinquency, they hoped that the new system would reduce recidivism. The undergraduate assistants served primarily as advocates for the child in dealing with others (e.g., the police and parents) and worked to set up behavioral contracts between each youth and important persons in his life.

Thirty-seven teenage girls and boys were referred to the program by the police when a court appearance for delinquency appeared imminent. The program was explained to the youths and their families, their rights as volunteer subjects were reviewed, and a signed consent form was obtained. Two-thirds of the youths were then randomly chosen to receive the experimental treatment—assignment of an advocate—and the others made up the control group who remained within the traditional rehabilitative system (police, courts, probation, and jail). There was a large decrease in repeat offenses among those in the experimental group, although there were no differences between experimental and control groups on personality tests. These encouraging findings were replicated in a second study. In addition, in the replication study, youths in the experimental group had significantly better school attendance.

With strong evidence from two studies for the efficacy of the advocacy intervention, Seidman and his colleagues felt that they could not justify exposing more youths to the control condition, the traditional court-probation system. Hence, in future research the beneficial effects of various youth diversion programs were compared with each other, not with a control group remaining in the standard criminal justice system. It should be noted that the experimental intervention in this study could influence the course of subjects' lives. When the repetitive offender cycle of the youths was broken, many were probably spared the fate of being labeled incorrigible delinquents and becoming habitual criminals. A somewhat similar experiment within the judicial system was conducted on the bail system in New York (Ares, Rankin, and Sturz 1973). Like large social policy experiments, local community change experiments raise the ethical question of whose goals and values are to be achieved. Are the goals of the scientist, of local officials, or of the target population most important? When does the scientist have the right to impose his goals and values on others?

EXPERIMENTAL THERAPIES

Psychotherapy and behavior modification techniques in general are not yet highly sophisticated or completely effective; much research will be required to perfect these aids to individual change. Research within the therapy setting has encountered many of the ethical dilemmas discussed above. For example, McConaghy (1969) attempted to treat homosexuals who sought to change their sexual orientation. He randomly assigned the men to one of three groups. The first group received electric shocks as they watched homosexual slides. The second aversion group received a chemical, apomorphine, as they viewed the homosexual pictures. The chemical produced an unpleasant nausea soon after it was injected. Another group comprised untreated controls. The subjects' sexual orientation and change was measured by subjective reports and by penile size changes in response to male and female slides. The two aversion groups showed a significantly larger change in sexual orientation than did the control group, but the cure rate was not impressive since only a small percentage of the homosexuals actually changed their sexual orientation. Note that, as in McConaghy's study, therapy researchers often do not have to confront the problem of whose goals are to be achieved because the client himself has sought aid in changing.

In McConaghy's study the control group received no therapy whatsoever. To provide service to this group, its members were also given one of the aversion therapies several weeks later. This allowed a test of the effectiveness of the therapy and still provided service to the clients. Another way of avoiding ethical problems related to no-treatment conditions is to give every group a different therapy and then compare the effects. This was the strategy used by Feldman and MacCulloch (1971) in their study on the treatment of homosexuals. Thirty homosexuals were randomly divided between three groups. The first two groups received different forms of shock aversion treatment. They watched slide pictures, and when a picture with a homosexual theme was shown, subjects would receive a painful electric shock, thus aversively conditioning them to the homosexual stimuli. Clothed and nude females were also shown, and no shocks were received for these pictures. Whenever possible, photographs of persons the client knew well were used. The third group received traditional verbal psychotherapy, the treatment most frequently used in clinical practice.

Feldman and MacCulloch discovered that 60 percent of those in the two aversive treatment groups showed improvement, whereas only 20 percent of those in the verbal therapy group decreased in homosexual orientation. These results persisted when measured in the one-year follow-up assessment. By comparing the effectiveness of different treatments, Feldman and MacCulloch did not need a control group that received no treatment at all. In fact, the verbal therapy comparison group received what was the treatment of choice of most therapists at that time. But another ethical concern remained: Were the therapists obligated to offer the most successful treatment to those in other groups after the study was over?

Ethical Issues
CONTROL GROUPS
Experiments in change include unique ethical problems associated with experimental and control group treatments. There are two major issues related to experimental treatment designs: the use of new treatments whose effects are unknown and hence potentially harmful, and the fact that some treatments may not be as beneficial as others. Thus we must ask whether it is justifiable to use different treatments for subjects in need when some treatments may be more harmful than others and some may be more beneficial. The ethical decisions related to different treatments should be based upon the relative value of the experimental treatment and current alternatives, both as predicted before the experiment and as established at the end of the study. (For a comparison of the issues related to medical experiments and control groups see Hill 1963; Katz 1972).

Strong new treatment. When there is good reason to believe, usually from other evidence, that an innovative treatment will be highly effective, then an untreated control group is often unethical, since one is knowingly exposing people to an inferior or even harmful condition when the experimenter is capable of offering a better one. Therefore a new treatment that is believed to be highly effective should not be contrasted to a no-treatment control but should be compared with the best treatment available at the time. This procedure is usually desirable on research grounds as well as on ethical grounds, because one usually wants to know not whether the new intervention is better than nothing, but whether it is superior to the current treatment of choice. For example, an educational researcher desiring to substitute multimedia sociology presentations for traditional lectures would not be justified in having a control class who received nothing and were simply given the tests and a grade in the course. Since the lecture format is the current method of choice of most professors, this treatment should be given to the comparison group. (The term "comparison group" is usually preferable to "control group" because it is more general and because some studies have several "control groups" or do not have a no-treatment control group.) The researchers would then be able to compare their innovative format with the standard lecture class.

There are cases when untreated controls or controls receiving an inferior treatment can be ethically used even though the experimental treatment is known to be strong. One of these exceptions is when the available resources do not permit treating everyone in need. This is an important exception because often social scientists cannot offer the innovative treatment to all who need the help. For example, in the negative income tax experiment described earlier, the control group who received no money was ethically justified because the experimenters obviously did not have the funds to assist all poor people. Similarly, in many therapeutic studies there are waiting lists for psychotherapy. Since limited resources prevent everyone's immediate entrance into therapy, these persons may often be ethically grouped together as an untreated control sample.

In an educational setting, Klaus and Gray (1968) and Gray (1971) offered an intensive preschool experience to deprived children. Equally poor children were randomly assigned to a no-treatment control group. After the experiment, IQ differences were found between the groups, and these persisted during the initial years of school. The randomly selected control group seemed to be placed at a disadvantage compared with the "head start" children. However, the untreated controls were like thousands of other disadvantaged children around the nation, and the investigators could not possibly help them all. In addition, the value of the preschool experience was uncertain before the study was conducted. And there could also be debate about the value of raising IQs and transfering middle-class values to poor children.

Another case when untreated controls may be employed even though the experimental treatment is believed to be very beneficial occurs when the potential subjects are not suffering from a serious problem and may be offered the treatment after the study is over if it proves effective. For example, Nay (1975) offered a training program to young mothers, teaching them child-rearing techniques that he had reason to believe would be very helpful. A no-treatment control group of mothers did not participate in the training, but they were offered individualized training after the experiment if they requested it. Thus, the no-treatment control group could be justified because the subjects were not really suffering from a serious problem without the training and because the treatment was available to everyone after the study.

The natural state argument. Sometimes the argument is advanced that untreated controls are ethical because they usually are not deprived of a benefit they already have, but are left in their initial state. Thus they have lost nothing because of their participation in the study. For example, Haynes et al. (1975) offered students a treatment for severe tension headaches that had lasted an average of five years. Virtually all the subjects had consulted physicians. A biofeedback method was compared with a systematic relaxation training approach. The treatments were designed to reduce or eliminate the tension headaches by teaching the participants to relax. Both techniques proved highly effective, almost eliminating the severe headaches. But a control condition was also included in which participants received no treatment, even though previous research had shown that biofeedback and relaxation were superior to no treatment.

According to the "untreated world" argument, Haynes had no obligation to treat the control group for their headaches after the study, even if he had the resources to do so, because these persons were simply left in their untreated state, suffering from headaches like thousands of others. However, there seem to be several weaknesses in this justification for untreated controls. One is that the necessity for untreated controls is questionable where one is comparing the effects of two treatments known to be highly beneficial. Another consideration is that untreated control subjects are often not like all those outside the experiment who are untreated, because the subjects have been led to expect help. For example, in the headache study, the subjects had answered an ad in the college newspaper. They expected to be helped with their problem and apparently had not consented to being placed in a control condition. Last, the researcher is often in one of the helping professions (e.g., a

137

psychotherapist) and may feel obligated to aid those who have assisted his research, or at least may feel obligated to refer them to a place where they can get help. Thus we see that the ethical researcher will usually feel an obligation to help untreated controls if the resources are available, irrespective of their condition if they had not entered the study. Even when various treatments are compared with each other, some researchers may desire to offer the most effective treatment to all participants after the study, especially if it was highly superior to the other treatments.

Treatments of weak or unknown efficacy. Frequently the effects of a new treatment may be entirely unknown or doubtful. The researcher may hope and believe the treatment will be effective but have little evidence to substantiate this. In these circumstances there is no problem with having an untreated control group, since no one knows whether they will be in a comparatively deficient condition. In this case the major issue is often whether the *experimental* subjects will be harmed or treated unfairly compared with those receiving no treatment or the treatments in current use.

If there is a currently used treatment that is somewhat effective, the researcher must have reason to believe the new treatment is superior or the experimental group will be harmed in comparison because it is not receiving the standard treatment. Often work with animals, analogue studies, and theory can be relied upon to suggest whether the new treatment will be equal or superior to the old. Take for example, the study by Paul (1966) on the treatment of phobias. The most widespread treatment for severe fears at that time was insight-oriented verbal psychotherapy. The investigator had reason to believe, based upon reports of case studies, that systematic desensitization (a relaxation-conditioning technique) might be more effective. The new approach also did not seem likely to be harmful. It was compared with the current "treatment of choice" and proved superior.

Of course where there is no currently effective treatment, there will be no issue of whether the experimental group will receive an inferior intervention as long as the new procedure is not harmful. Langer and Rodin (1976), for example, wanted to aid elderly residents of a nursing home. The researchers believed that the debilitated condition of many senior citizens in "old folks' " homes occurs partly because they have no responsibilities and make few decisions. Of course there was no current "treatment" for this decision-free condition. Langer and Rodin gave the residents on one floor some small responsibilities—for example, caring for a plant, deciding when the movie should be shown, and arranging their own rooms. On another floor, the elderly in the control group were "taken care of" in the usual fashion. During the three-week test period 93 percent of those in the experimental group were rated by the nurses as improving, whereas 71 percent of those in the control group actually became more debilitated. An innovation may thus prove to be a boon compared with the condition where nothing is being done.

For some innovations there may arise the question whether the new intervention could injure people. For example, Fo and O'Donnell (1975) created a community-based intervention system for delinquents. Delinquents and potential delinquents referred to the program were matched with local nonprofessional "buddies" from their neighborhoods. Whereas those with recent criminal histories were helped by the program, potential delinquents who had no criminal record committed *more* major offenses in the ensuing year than did untreated controls. There are, unfortunately, some cases where harm cannot be foreseen. In those cases where possible harm can be predicted, informed consent is very important.

It is important to note that researchers have a *positive obligation* to do experiments on new techniques during the early stages of innovation. It is precisely at this time that the most rigorous experiments can be done, because untreated controls can often be used without problems. Also, once the intervention becomes institutionalized or highly popular, it will be harder for research evidence to counter its use. Last, early controlled experiments are important because they will help prevent the widespread use of worthless treatments. Thus it may often be unethical to use unproved treatments without evaluating their effectiveness.

Summary. There are several considerations in the use of experimental multitreatment designs. The various treatments should be compared a priori, and if one is *definitely* superior it will be hard to justify the experimental use of inferior treatments (or no treatment) on comparison groups. If the treatments (or no treatment) all seem potentially to be of similar value, then they may be compared experimentally. The experimenter should consider whether it will be feasible to expose all subjects to the superior treatment after the study. There are some cases when an inferior-treatment or no-treatment group may be included even if the experimental treatment will probably be more effective. One frequently met case is where the scientist cannot offer the superior treatment to everyone who needs it because of lack of resources or other factors not under his control. Another case is when a no-treatment group is given the treatment after the study (but does not suffer substantially in the interim). Experimental designs are important because social scientists have an obligation to rigorously evaluate the effectiveness of innovative interventions from an early stage in the program. (See M. S. Frankel 1976 for a further discussion of the ethics of experimental and control groups.)

INFORMED CONSENT

Informed consent is important in experimental intervention programs because the person or group may be changed, and perhaps changed permanently, by the research experience. On the other hand, the targets of an experimental treatment may benefit greatly, much more than is typical in purely theoretical research. The magnified importance of voluntariness when one may be permanently changed by an experiment prompted Donald Campbell (1976, p. 16) to write: "All participants in an experimental program should be informed in advance of all features of the treatment and measurement process which they will be experiencing which would be likely to affect whether or not they volunteer for the program, or how they would manage their lives during the program. Institutional Review Boards should be provided the specific wordings of the information provided potential participants when seeking their consent."

Information. As was stated in chapter 3, there are usually several key elements of information that should be communicated during the informed consent procedures: the subject's rights and obligations; that he can withdraw at any time; the procedures the subject will undergo; the risks; and the potential benefits. The subject should certainly be informed if the new treatment could be injurious. If the study includes deception, the ethical problems become pronounced if the subject is thus prevented from knowing about risks or about what procedures will take place.

One question often asked about information to be given before a change study is: Must subjects be told about the other conditions or treatments? For example, in the income-maintenance study, should low-support subjects have been told about higher-income conditions? The answer is that subjects need be informed only of the risks and procedures to which *they* will be exposed. If the random assignment to conditions has already been made when subjects are approached, they can be told about that condition and given a choice to participate or not. They usually need not be informed about the other conditions since participating in another treatment condition is not an option for them and will not affect their welfare. Telling subjects about other conditions could sometimes seriously impair the value of the study. In the negative income tax study, for example, participants exercised fully informed consent for their own family's condition but were not told about other conditions (Kershaw 1975). It is likely that knowing about higher-income groups would have caused many not to participate, hence jeopardizing the initial equivalence of the groups. Subjects were effectively protected since they were not directly exposed to the effects of other conditions they were unaware of. Subjects may react negatively if they later learn of other conditions in an experiment after it has begun (Wortman, Hendricks, and Hillis 1976), and this reaction may be considered in designing the study.

Whenever subjects enter an experiment and may be exposed to any of the treatments or to no treatment (when prior assignment has not been made), they should be told at least generally about the possible conditions to which they might be assigned. For example, in a study on the treatment of reading deficiencies, the parents should be told in general about each of the conditions if assignment has not yet been made so they will realize that their child may not be exposed to the innovative intervention. Parents of children who are assigned to the control group might be offered the treatment for their child after the study if resources permit.

Freedom of choice. Participation in the treatment is normally voluntary. This does not mean that the subject can choose which condition he prefers but means he is given a choice whether or not to participate at all. However, there have been many cases where subjects have not been given a choice whether to participate in change studies, for example, when the subject does not even know he is in a study. In one study 2,139 unknowing high-school students were used as experimental subjects in their usual courses (Page 1958). Different types of teacher comments were randomly written on papers that were being returned. Some students received encouraging feedback, others received none. The dependent variable was the student's performance on the next test, and those who received positive comments performed significantly higher than those who did not. Notice that teacher comments are part of everyday school life, so that the experimental treatments were unlikely to have a profound or lasting effect on pupils. However, the random assignment of feedback conflicted with the teachers' basic function of giving students help, including accurate feedback.

Compare the teacher-comment experiment with a study by Cook (1970) in which subjects also did not know they were participating. In this study subjects were presented with a treatment that they would be less likely to encounter in everyday life and that was designed to permanently lessen their racial prejudice. Subjects were hired for a part-time "job" two hours a day for a month. The extremely prejudiced subjects were unknowingly exposed to a highly coordinated effort to improve their attitudes toward blacks by having them work side by side with a black woman who was highly personable, competent, and ambitious. A number of the experimental subjects decreased greatly in prejudice. Although most would agree that a reduction in prejudice is a good thing, the subjects might not have agreed to this before the study. They did not exercise informed consent or even know they were subjects. To be less prejudiced may even have caused them difficulty if their families and friends were still prejudiced. Are such extensive change experiments ethical without informed consent? Certainly such experiments should not be attempted if they expose subjects to substantial risk. They should always be reviewed by others including representatives of the subject group. Ethical and values issues must be carefully weighed before a scientist embarks on a direct change program without subjects' knowledge or consent.

Sometimes change attempts are conducted by society and the research occurs secondarily. In these cases society may force the change program on subjects and the scientist's role is simply to measure change. For example, prisoners and delinquents are exposed to many change attempts by the judicial system. They may be placed in jail, in a halfway house, or on probation, and their consent is rarely considered. When the government or some other institution implements an experimental program, scientists may participate in assessment yet not be ethically responsible for the change attempt itself.

Criminal offenders must usually be considered separately from other research subjects because they have already lost considerable freedom. For example, in the adolescent diversion project the delinquents would have had little choice about their disposition anyway. Similarly, people in educational settings have usually lost certain freedoms. They must take tests at certain times, write papers, and so forth. The teacher usually plans the educational program without consulting the students. Hence it is often helpful for ethical purposes to distinguish between the change program itself and the assessment research. The subject may be exposed to the change program because others (for example teachers and the court) hold power over him, and informed consent may not be necessary (or even relevant) for the intervention per se. The change program may not reduce the subjects' freedom since they may have little freedom anyway. However, it is often still possible and desirable to gain consent for the measurement aspect of the research even if the subject has no choice about entering the change program.

Summary. Informed consent is particularly desirable in intervention studies because people may be strongly influenced by the research. It is important to inform potential subjects about the procedures, risks, and benefits, and about their right to withdraw. If he is assigned to an experimental condition before his consent is obtained, the subject does not need to learn about the other conditions. When a subject may receive any of the treatments, he must learn of the risks inherent in all conditions and usually should know whether some subjects will be receiving a placebo or no treatment. A researcher who contemplates conducting an experimental intervention program without obtaining subjects' informed consent must consider this decision carefully. In this case subjects should not be exposed to substantial risk and a review panel should carefully examine the study. Even then, studies that may change subjects without their consent or knowledge often will not be ethical. Sometimes research programs may be carried out without informed consent when agencies have already withdrawn the subjects' freedom and a change program is being imposed by society (e.g., the courts).

VALUES AND CHANGE

Research endeavors inevitably reflect to some extent the values of the researcher. (See chapter 12 for a more complete discussion of this issue.) Questions about values are important in intervention research because change attempts reflect strong value commitments (Warwick and Kelman 1973). The research examples described in this chapter were attempts to change patterns the scientists believed were deficient or wrong. Of course what one defines as needing to be changed is a direct reflection of what one believes to be ideal, and hence reflects one's values and goals. Take for example, research in improving classroom learning. The researcher obviously values the type of learning that occurs in schools (Gray 1971). In therapy research, too, there are ideals of adjustment or self-actualization that reflect the therapist's values. The psychologist who tries to "cure" a homosexual or an overaggressive child is certainly acting according to his own values. Even when the homosexual or the parents of the child have requested the psychologist's aid, the change ordinarily must not conflict radically with the therapist's values or he would not be likely to cooperate. For example, a therapist would be unlikely to help a person become more vicious or become a pedophiliac.

Values enter into experimental intervention programs at several stages: belief that there is a problem, definition of the problem, and deciding the target of change. For example, when ghetto children have difficulty learning to read, a social scientist may seek to remedy this "problem." Note that the scientist values reading and considers inability to read a deficiency. Note too that the basic problem may be conceptualized in different ways (Seidman 1977; Warwick and Kelman 1973). The problem may be seen as an attribute of a school system not attuned to the needs of the ghetto child. Or the scientist may fault the child, not the institution. The child's dialect, family environment, achievement motivation, diet, or genes may be seen as the root of the problem. Which of these is believed to be the basic cause will determine the type of change program instituted.

Because values inevitably enter into social intervention processes,

144

it is necessary for the scientist to take special precautions to remain objective. However, from an ethical standpoint there is another issue: Whose values determine the direction of the change program? In a pluralistic society, the change agent, the research scientist, the target population, and societal officials may have different goals and values and thus may differ about the best solution to the problem. Indeed, what may be a "solution" for one group may be considered exploitative or worthless by another group. Take, for example, the solution to urban riots. Public officials are likely to see riots as a breakdown in social order and possibly conceive of the solution as a larger police force and more social control. Many ghetto residents will believe the riots result from a breakdown of social justice and will see the solution in more jobs and a more equitable distribution of power. The solution of one group may not be appealing to the other because they see the basic problem differently.

The least questionable experimental change programs occur when the scientist, the target population, and societal authorities all agree on the definition and possible solution of the problem. For example, if a new reading program is tested in a local first grade, the program may or may not prove successful. But it is probable that the experimenter, the school officials, and the parents all agree that better reading ability is a valuable goal. Similarly, many clients in psychotherapy and many parents of children in innovative preschool programs believe the goals of the programs are good ones.

When there is disagreement about the goals to be obtained, ethical questions arise if the recipients of the change attempt do not believe the goals are appropriate. For example, if a homosexual wants to remain a homosexual, the change agent faces a real ethical problem in instituting a program to change the person. The researcher can probably justify attempts to change someone against his own wishes on some occasions, as when a mental patient is a danger to himself or others. But intervention attempts that are not desired by the target population should be carried out cautiously. Is there widespread agreement that the change will be beneficial, or do only some segments of society agree? Is the problem defined mainly in terms of the researcher's values and biases, or have other perspectives also been considered? Ordinarily the target population should agree about the existence of the problem and the possible solution recommended. Frequently it is helpful to get their input when the change program is planned, especially in community change efforts. The less consensus there is on the need for change or the desirability of the type of change proposed, the more input the researcher should gain from the subject population and the more carefully he should examine the research goals.

Campbell (1976) points out that many recent social science experiments fall under the rubric of "blaming the victim" (Ryan 1971) because the researcher attributes the problem to the victims of mistreatment. For example, researchers often emphasize the inadequacies of poor people rather than the societal structure that may cause poverty. Poverty is attributed to lack of ambition and so forth rather than to unfair employment, inadequate schools, and such. Thus the victim is typically blamed for his own problems. To avoid blaming victims one must take their perspective into account and obtain more input from the recipient population when formulating research programs (Seidman 1977). Sometimes this input will focus the intervention program on businessmen or government officials rather than on the "victims." At times a community may request aid in changing—for example, through education. At other times they may ask for help in changing external blocks to progress. Researchers should be able to formulate intervention attempts from the perspective of those not in power, so that the research enterprise will reflect the interests of all segments of a pluralistic society. Though the scientist may be the expert on change, he should try to shape his goals according to input from the society, especially from those most directly affected.

Conclusion

Intervention research is aimed at studying programs that try to change individuals, groups, or society. Research of this type raises the usual questions about privacy, deception, and harm. In this chapter we have assumed that the scientist is actively involved with the change program. Where the scientist is merely evaluating a program instituted by others, some ethical considerations may not apply. But when the scientist has a more active role in the intervention program, he must confront ethical issues related to the use of control groups, informed consent, and the value basis of the change program. Control or comparison groups raise the question whether some experimental groups may be harmed or at least at a disadvantage compared with other groups. Usually the scientist should make sure that all groups are treated fairly, and he may offer the most beneficial treatment to all groups after the study. Usually it is not ethical to knowingly place some subjects at a serious disadvantage unless one remedies this later or unless one has too few resources to offer the effective treatment to all those in need.

Informed consent normally is a must in intervention research because the subjects may be permanently changed by the experience. However, there are situations in which society imposes the change program and there is little or no opportunity for informed consent. But it is still desirable to gain individual consent for the research evaluation part of the program.

The question of values is very important in programs to change people or societal institutions. When all persons and agencies concerned with a problem agree on a goal and a common set of solutions, few ethical problems arise. But if the recipients of change do not agree, ethical problems arise. The scientist thus should seek input from many sources, including the recipients of the change attempt. This process can be difficult and may not yield unambiguous answers (Warwick and Kelman 1973), but scientists need a broader outlook than one that reflects only middle-class values or the goals of the most powerful segments of society.

For additional reading see Campbell 1969*b*, 1976; *Evaluation*; *Evaluation Quarterly*; M. S. Frankel 1976; Kershaw 1975; MacRae 1976; Riecken 1975; Riecken and Boruch 1974; Rivlin and Timpane 1975; Warwick and Kelman 1973.

III   Professional Issues

9 Honesty and Accuracy

As long as my breath is in me and the Spirit of God is in my
nostrils, my lips will not speak falsehood and my tongue will
not utter deceit.

> The Book of Job

After it became obvious how tedious it was to write down
numbers on pieces of paper which didn't even fulfill one's own
sense of reality and which did not remind one of the goals of
the project, we all in little ways started avoiding our work and
cheating on the project. . . . It began innocently enough, but
soon boomeranged into a full cheating syndrome, where we
would fake observations for some time slot which we never
observed.

> Anonymous Research Assistant
> From Roth (1966)

Following the model of Job, the ideal scientist should be completely
accurate and honest in conducting and reporting research. The most
basic rule governing scientific work stipulates that research must be
conducted objectively and reported honestly. But the numerous
cases of dishonesty by experimenters and research assistants cited in
this chapter reveal that dishonesty in research may not be as rare as
is commonly believed. We hope that by reviewing the motivations
underlying cheating and the destructive effects false results have on
science, we can make it clear why ethical research requires complete
scientific honesty.

The strength of scientific knowledge is that it relies upon observation, not merely opinion, to discover truths about the natural world.
Falsifying data destroys people's faith in scientific findings because
it eliminates the accurate observation upon which scientific progress
is built. In addition to observing accurately, scientists must communicate their findings honestly. Scientific advance usually rests on
the work of many investigators. Since the events studied are so

complex, no single researcher can hope to go very far alone in describing and understanding the world. The accumulation of accurate scientific knowledge is so time-consuming that it may take many human lifetimes to build a foundation in just one area. It is thus essential to scientific progress that findings be made public so that researchers can build upon the discoveries of others. Perhaps more than any other human enterprise, science relies on cumulative knowledge, and no progress can be made without the sharing of accurate scientific information.

If honesty is so important, why do some scientists and research assistants contaminate scientific knowledge by using biased procedures or by altering or faking their data? Despite the importance of accuracy, there is often a temptation to be dishonest because of strong motivations to produce significant results. Finding and publishing significant results will partly determine employability, salary, promotions, and national reputation. Without impressive research findings, the young researcher may lose his job, the older scientist may not be promoted. The policy of "publish or perish" and the usual practice of journals to publish only positive results can lead scientists to present their findings as stronger than they really believe they are. In addition, the scientist may be attempting to support his own theory and will thus hope the results turn out in the predicted way. Usually he has invested much time and effort in the study and will be disappointed if the data are inconclusive or not publishable.

Sociologists of science have described how the structure of the scientific enterprise in the West pressures scientists to achieve recognition (Barber 1962; Barber et al. 1973; Merton 1957a, b). It is usually through peer recognition for original scientific work that the scientist learns that his work is meritorious. Recognition by peers comes for *original* scientific work on important problems; therefore there is intense pressure to be first to publish a particular discovery. Since paradigms tend to make certain problems seem the most important, many scientists are often working competitively in an area, all hoping to publish first. The pressures of competition and the need for peer recognition tend to set up a situation in which misbehavior can be highly rewarding. In such a system, outright fabrication will occur occasionally, and lesser transgressions of the mores of science may occur frequently. Thus we can see that the motivations to falsify results must be overcome by a strong commitment to honesty.

152

Although scientists are usually committed to truth despite pressures to produce certain results, their beliefs and scientific paradigms often bias their results from the outset because they determine what questions will be asked, what data will be collected, and how the findings will be interpreted. This basic type of bias is partially corrected because other scientists may work from different perspectives and criticize those approaching problems in an alternative way. This type of paradigm bias can also be reduced if a scientist painstakingly tries to see different viewpoints and examines his own research critically (see also chapter 12). The bias introduced by the investigator's paradigm (T. X. Barber 1973, 1976) is not considered unethical within scientific circles, but it can nevertheless distort findings and lead an investigator to other questionable ethical practices such as citing only the studies that support his viewpoint, reporting only supportive data (Dunnette 1966), and repeating data analyses until one turns out in the way desired.

Whenever scientists depend on an experiment to produce specific results, they may be tempted to bias the experimental procedures or even to falsify data. To counteract this temptation, Platt (1964) has recommended a "strong inference approach" to research. According to Platt, researchers should pit differing theories or hypotheses against one another within the same study. Thus the researcher is not dependent on the data's coming out in one specific way and the study is more likely to be publishable even if it does not have the results the experimenter hoped for.

Research assistants may be tempted to falsify results during data collection and analysis for a variety of reasons. They may hope for praise or a better grade from their supervisor. In some cases, the motivation may even be altruistic: a desire to help the scientist they are working with obtain the results he wants. Sometimes research assistants cheat because they do not realize the importance of standardized procedures. At other times they may falsify results because they are tired of collecting data and begin to take shortcuts or manufacture data that are troublesome to collect. These and other motivations often tempt research assistants to modify standard experimental procedures, fabricate data, or alter findings. Dishonesty by research assistants is just as destructive to science as dishonesty by scientists. Like the scientist, the research assistant must make a personal commitment to accuracy and honesty, realizing that temptations to fudge data will arise but must be

153

resisted. In the following section we review cases of outright fraud and the negative consequences that follow such practices.

Outright Falsification

There are cases in both the natural and the social sciences in which researchers have been caught falsifying data. A famous case was that of the Austrian biologist Paul Kammerer who obtained research results that supported the inheritance of acquired characteristics (Koestler 1971). Kammerer found that salamanders raised on a yellow or black background developed more of this color in their own pigmentation. Furthermore, the eminent biologist found that this acquired characteristic was passed on to offspring. However, after about twenty years, during which time Kammerer was finding similar results, an independent scientist examined one of his specimens and found that it had been colored with ink! When it was publicized that the specimen had been tampered with, Kammerer was disgraced and committed suicide. It is quite possible that a research assistant or a jealous colleague, not Kammerer, was responsible for the fabrication. Another well-known scientific scandal was the Piltdown hoax, which threw physical anthropology into confusion for a number of years (see chapter 1).

More recently, cases of fabrication within the natural sciences (e.g., DuShane 1964; Lappe 1975; Rensberger 1977) and social sciences (Asher 1974; Wade 1976) have been made public. In chapter 1 we described the Levy case, in which a bright young scientist was caught altering the results of extrasensory perception research. Despite Levy's immediate resignation, his falsification of data will harm research on extrasensory perception by increasing doubt about all positive findings, including those of honest researchers.

Within psychology a possible hoax has recently emerged in the controversial area of the inheritance of intelligence (Evans 1976; Gillie 1977; Panati and MacPherson 1976; Wade 1976). One reason the heritability of intelligence has been so hotly debated is that the answer *may* have implications for whether racial differences in IQ are inherited or due to environmental differences. A leading proponent of the inherited intelligence viewpoint was Sir Cyril Burt, a British psychologist who was knighted for his scientific work. Burt claimed to have measured the IQs of more than fifty sets of identical twins who were separated early in life. The correlation between the separated twins' IQs was high, suggesting that IQ is largely genetic, not environmental, since the twins were purportedly living in very different environments. These and other of Burt's data have held a central place in supporting the hereditarian position. For his important work, Burt received a coveted scientific award of the American Psychological Association. But since his death, evidence has been uncovered that some or most of Burt's data may have been faked. It appears that Burt was so convinced that his opinion was correct that he simply considered data a way of showing others what he already felt sure was true. Burt apparently made up coauthors for his writings, supplied data for twins who were never tested, and reported results found in one study in subsequent studies so that there appeared to be perfect agreement between them. Burt also may have fabricated results to negate the arguments of critics. It is important to note that the influence of Burt's data was not confined to academic circles—the entire British school system of tiers was based largely on Burt's belief in fixed intelligence. For defenses of Burt see Eysenck (1977) and Jensen (1977).

Although cases of outright fraud by experienced scientists *may* be rare, there is evidence that falsification by young scientists and research assistants is fairly commonplace (see Azrin et al. 1961; T. X. Barber 1973; Hyman et al. 1954; Rosenthal 1966; Roth 1966). For example, Azrin et al. (1961) asked graduate students to complete a field experiment on verbal conditioning, a replication of an earlier published study (Verplanck 1955). In the course of conversations student experimenters were to "reinforce" or "extinguish" opinions stated by the other person by agreeing with them (reinforcement) or remaining silent (extinction). In fact, the study turned out to be impossible to carry out by the specified procedures. Despite the impossibility of running the study in the prescribed way, fifteen out of sixteen graduate students reported to their professor that they had successfully run the study. Only one honest graduate student reported that he found it impossible to conduct the research in the prescribed way. After his report, eight of the other students admitted they had deviated significantly from the prescribed procedure in collecting the data.

Azrin et al. replicated the verbal conditioning study, again using undergraduate students to collect the data. One of the students who was working undercover for the professors talked to the others about how they were doing the study, stating that he was having trouble running his experiment. Twelve of the nineteen students approached, 63 percent, told him they had faked part or all of their data. Five more told him that they had simply changed the prescribed procedure to make it work. Finally, when the investigation was replicated by the four experienced scientists, they could not complete the study according to the proper procedure. In light of the findings of Azrin et al. and the fact that Verplanck's data were collected by students, it is probable that the original report was based on fabricated data. Other reports of cheating in the social sciences can be found in Wyatt and Campbell (1950), Sheatsley (1947), Guest (1947), Roth (1966), and Rosenthal and Lawson (1964). These findings suggest that falsification of data may be more widespread than is usually believed, particularly among students and hired assistants. If this is true, a sizable proportion of published findings may be based on quicksand. Scientists and research assistants alike must understand that their work is fundmental to scientific progress and that dishonesty will completely invalidate it.

Sometimes experimenters start out with the best intentions and then become sloppy or make up difficult data as they become bored with running a large number of subjects (e.g., Roth 1966). Often researchers assume that falsified data from one subject will not really matter since it will be added to a large pool of accurate data. But when this occurs they often falsify data for an increasing number of subjects as the experiment progresses. A large part of the data thus is partly or wholly falsified, so that the entire pool is inaccurate. For example, Hyman et al. (1954) reported a case in which hired interviewers faked interview results. Each employee interviewed people planted by the experimenter, who were deliberately difficult to interview. The proceedings were secretly tape-recorded, and it was found that one-fourth of the interviewers fabricated most of the data for these cases. All of them "fudged" some of their data. It is often tempting to do this when one thinks one knows what the person will say, or when the questions to be asked are somewhat embarrassing. Scientists and research assistants must realize that "dirty data" is worse than no data at all, since people will accept it as objective. Fabricated data are nothing more than a guess at how the data might have turned out if the study had been conducted correctly.

We have seen that being caught falsifying research results can lead to personal disgrace and to expulsion from the scientific

156

community. The data collected are worthless. Even where the results are not published and the cheater is not found out, the effects of falsification can be harmful, as the following example illustrates. Some years ago an undergraduate student we knew conducted a study on arousal and exploratory behavior using laboratory rats. The student ran several rats in each of the experimental conditions, and a trend seemed to appear in the data. The student had much studying to do for other classes, so he made up the data for many more rats without actually running them, a procedure sometimes called "dry labbing." The research report was interesting and the professor encouraged the student to submit it for publication, especially because the results contradicted those found in a similar published study. Fortunately, the student's conscience caught up with him at this point and he did not submit the study, offering the excuse that some of the procedures had not been followed carefully enough. The professor, encouraged by the findings, himself tried to replicate the study but could not do so. Even though the study was not published, the professor wasted a large amount of time trying to replicate the falsified results. Dry labbing experiments in undergraduate laboratory courses may be commonplace; admittedly, this form of cheating does not have as destructive consequences as faking data that will be published. But dry labbing should be avoided because it will probably increase the chances that the young scientist will fake or alter future data that may be published. Also, if the student is caught he will probably be prevented from pursuing a professional career and may be expelled from the university.

Student researchers might rationalize falsifying data by saying they know how it will come out. Therefore they can save themselves time and energy by making it up. However, since experienced social scientists have found that they are usually unable to guess how a study will come out, surely the guesses of novices will be inaccurate. A dramatic illustration of our inability to guess about research findings was offered by Milgram (1963). In Milgram's study an experimenter commanded a subject to give progressively higher levels of shock to another "subject" (a confederate). Most subjects in the actual experiment administered high levels of shock, even after the confederate appeared to be dead or unconscious. Amazed at the outcome, Milgram wanted to find out what others would predict. When he asked Yale psychology students, they guessed that 3 percent or fewer of subjects would give the highest shock (65 percent actually did so). When psychologists and psychiatrists were queried, they were similarly inaccurate in their predictions.

Social scientists are often surprised when, after formulating hypotheses based on theories and "common sense," their data prove them wrong. In fact, it is likely that social scientists can accurately predict in detail the outcome of less than half their studies. How then can we hope to fake results accurately? The answer is simple: We cannot. Science differs from other sources of knowledge such as common sense, intuition, and opinion precisely in that it is based on objective observation. When results are fabricated, they are nothing more than personal opinion, in which there is notorious disagreement and inaccuracy. Opinions presented as opinions are acceptable, but opinions should never be presented as facts.

One reason most scientists resist temptations to fabricate results outright is that they realize the consequences. If caught, they will be ostracized from the profession. And in any case, they will have contaminated scientific knowledge with sham results that may mislead others. Scientific progress depends upon researchers' willingness to forego promotions and other incentives in order to remain scrupulously honest. Social scientists often conduct studies that do not produce publishable data or collect data that do not support their theories. Progess in science depends on such individuals' not altering or falsifying data to suit their personal needs.

We have known student research assistants who have invented some or all of the data for a study, and other students who have faithfully collected all the data but then altered it to obtain more impressive results. Usually these students did not realize the potential ill-effects of this falsification. They seemed to consider it comparable to cheating on classroom tests, a common practice in colleges (Smith, Wheeler, and Diener 1975). However, falsifying scientific results is far worse than cheating on a test, which has its greatest effect on the cheater himself. Faking results may benefit the individual in the short run; but it negatively affects the entire scientific enterprise.

Student research assistants should be educated about the very detrimental effect falsification has on the scientific process as well as on their own professional future. In addition, it is desirable to build safeguards into the data collection procedure to prevent falsification by assistants, thus helping them resist the temptation. For example, the data collection can be periodically checked, two people can independently code all data, and frequently the assistants can be ignorant of the hypothesis.

Biased Results

Scientific objectivity can be destroyed by less blatant practices than outright falsification. Data can be deliberately biased by the way they are collected, thus degrading the scientific method. For example, one investigation of the financial status of the elderly was supported by the American Medical Association (Science and politics 1960). Old people were interviewed about their income and savings, and this knowledge was to be used to determine whether Medicare was really necessary or whether older citizens could afford medical care on their own. But the study was biased from the outset by excluding certain groups. Not interviewed were: minority groups, those on welfare, those in institutions, and those living in apartments. That is, only whites who lived in houses and were not on relief were interviewed—precisely those who would have by far the most money. Not surprisingly, the study "revealed" that the elderly are not poor and do not need Medicare. The data were obviously biased by the sample used and tell us little about the financial status of the average senior citizen.

A study can be biased by how the sample is selected, by how questions are asked, by subtle messages conveyed to subjects, and in many other ways (T. X. Barber 1973). Social scientists should be trained in how to overcome these sources of error and should work to make their research as unbiased as possible and to overcome even subtle, unintentional bias effects (T. X. Barber 1973; Rosenthal 1966; Rosenthal and Rosnow 1969). A number of techniques have been developed to minimize experimenter bias; for example, raters may be "blind" as to which condition each subject is in so that their ratings remain as objective as possible.

In laboratory experiments and survey research a formal "protocol" designed specifically to eliminate biases usually specifies exactly what is to be done and said in each of the conditions. It is important that the protocol be followed conscientiously; otherwise the careless investigator will introduce uncontrolled sources of error into the findings. Research assistants may introduce error because they do not understand the importance of maintaining highly standardized procedures. The more standardized and the better rehearsed the research protocol is, the fewer biasing influences can creep in. Since

159

most people will inevitably alter the protocol slightly to fit their personal "style" (Friedman 1967), there should be monitoring of how closely the procedures are being followed and how much variance there is in the protocol from subject to subject. Feldman, Hyman, and Hart (1951) have shown that a loose experimental protocol can produce unreliable data.

In laboratory studies and field surveys the investigator can distort findings by the way he gives instructions or asks questions (Friedman 1967). By stressing certain words or through certain facial expressions, an experimenter may bias subjects' responses. In a laboratory experiment the experimenter can lead subjects to respond differently by giving the instructions in a different tone of voice. Needless to say, it is unethical to try to make the experiment "come out right" by introducing such gimmicks if they are not part of the phenomenon to be tested. It is often impossible to eliminate all bias in research, but the scientist's job is to strive for the ideal of complete objectivity. Where he is aware of possible biasing influences he attempts to counteract them, and he reports possible sources of bias where they cannot be circumvented.

Distorted Reporting of Findings

A social scientist may carry out a study objectively and yet miscommunicate the findings. One clear situation is when the investigator makes "slight" errors in recording the data so that the findings will better fit his expectations. But there are other unethical practices that may not jar the conscience so severely. An investigator may complete several studies whose results do not substantiate his hypothesis and finally report only the one study that does support his theory. A whole study may be suppressed because the findings are unpopular. McNemar (1960) suggests that some researchers simply discard findings as "bad data" if they do not support their hypotheses. Sometimes an investigator collects large amounts of data in a single study and selectively reports only those that confirm his theory, not mentioning contradictory or nonsupportive findings. Perhaps a more common practice, though also ethically questionable, is conducting a large number of statistical tests on the results and reporting the few significant analyses as though they were the only ones computed. One is responsible to report or mention all data that bear on important aspects of the research. Once the full results are reported publicly, the investigator is free to argue for a particular interpretation; other scientists may then consider his arguments in light of the procedures used and the results found and draw their own conclusions.

Yet another method of biasing results is by making errors in analyzing or reporting the data. For instance, experimental results on which inferential statistics are completed may yield a probability value of $p < .07$, but the investigator may report the result as significant at $p < .05$ since it was "so close." Indeed, Wolins (1962) discovered that there are many such errors, intentional or unintentional, in published journal articles. Wolins requested original data from thirty-seven studies from the authors of the articles. Twenty-one of these investigators reported that the data had been misplaced or destroyed. Of the seven published studies for which data were received and reanalyzed, three included large errors in statistical analysis that altered the conclusions drawn from the data. If this sample is at all representative of the percentage of error in the scientific journals, it is terrible news for scientific progress. Dunnette (1966) has also highlighted the problems of questionable research and presentation standards, and Duce (1975) and others have pointed out that references are often cited inaccurately. Many errors are probably due to carelessness or ignorance, not to deceit, but inadvertent errors can be just as detrimental to scientific progress as deliberate distortions. Hence every researcher is obliged to analyze data carefully and to consult experts where his statistical knowledge is limited. If a researcher discovers a significant error in a report he has published earlier, he should bring the mistake to the attention of the journal editor, who can decide whether a printed correction is desirable. If the error is likely to mislead others about important aspects of the findings, an effort should be made to publish a correction.

Summary

The costs of falsifying or distorting scientific findings are high. Falsified results are worthless and ultimately may lead to questioning of the entire enterprise. Unfortunately there is evidence that outright falsification and deliberate biasing of results do occur. Less blatant forms of dishonesty such as reporting only supportive findings and slightly altering statistics are probably fairly common. The scientist should anticipate temptations to alter data and be prepared to remain honest at any cost. Cheating not only is detrimental to science but will be disastrous for the individual who is caught at it. More subtle forms of biasing like the "investigator paradigm effect" (T.X. Barber 1973) can also detract from scientific advancement, and each investigator should critically examine his own work to reduce this effect.

There is simply no ethical alternative to being as nonbiased, accurate, and honest as is humanly possible in all phases of research. In planning, conducting, analyzing, and reporting his work the scientist should strive for accuracy, and whenever possible, methodological controls should be built in to help experimenters and assistants remain honest. Biases that cannot be controlled should be discussed in the written report. Where the data only partly support the predictions, the report should contain enough data to let readers draw their own conclusions. As will be discussed in chapter 12, it is impossible for a scientist to be completely objective; but if he reports his methods and results accurately, readers can evaluate the data for themselves and control for the earlier biases if they replicate the work.

For a further discussion of honesty in social science research see T. X. Barber 1973, 1976.

10 Publishing and Grants

A senior psychology student takes an independent study course from a young professor, and together they plan a research study. The student reviews the literature in an area assigned by the professor, and based upon this review the scientist designs a clever experiment. The student then collects the data, analyzes them, and writes up a rough draft of a complete paper. Since she hopes to attend graduate school, the student asks if she will receive second authorship on the paper. The professor says he will have to decide and eventually publishes the paper under his own name only, with a footnote thanking the student for her help. The student, now in graduate school, sees the paper and feels she deserved second authorship. But since she does not know the standard procedure in authorship determination, she decides to forget the matter.

The student in this example was faced by a question of professional ethics: How is authorship of a paper fairly determined? Professional ethics not only directly influence our treatment of research participants; they dictate fair practices between scientists themselves. In this chapter, we shall discuss two issues important to the work and advancement of a scientist: publishing and monetary support for research.

Publishing

Since collecting and disseminating information is a primary goal of scientific research, publishing reports for the scientific community is a central task for social scientists. Despite the importance of authorship for social scientists and for the advancement of science, little has been written on the guidelines that govern publication credit. Yet publication issues often come before professional ethics committees in sociology (Epstein 1975), psychology (Seashore 1976), and other disciplines.

DETERMINING AUTHORSHIP

Personal recognition for scientific work usually comes as authorship of published articles. Published work can increase the chances of being admitted to graduate school, getting a job, and receiving promotions and salary increases. For example, Tuckman and Leakey (1975) estimated the value of the first publication for an assistant professor of economics at $12,340 by comparing projected salaries over the course of a career for individuals with and without publications. They did not include some sources of income such as consulting or lecture fees that are likely to be increased by publication. Clearly, published research is extremely important to the careers of research scientists both financially and in terms of professional reputation.

The amount of credit a person receives for scientific work usually depends on the order of authorship. In the social sciences the greatest recognition is usually given the first or "senior" author; so he should be the person who has made the largest *scientific* contribution to the study. If the contributions are equal, the first author can be decided randomly, such as by the flip of a coin. Another alternative is to use alphabetical order of names and indicate this in a footnote. Where a team of investigators work together and make equal contributions to a series of studies, senior authorship can be rotated.

"Junior authors" are all other investigators listed as authors, and they also should have made significant scientific contributions to the study. Those who contribute to the study in ways that have no direct influence on its design or content usually deserve credit in a footnote. Although this type of credit is not as helpful for career advancement as authorship, it is appropriate for nonscientific contributions. For example, lending scientific apparatus and routinely running subjects usually deserve footnote credit, not authorship.

The general principle governing publication credit is that authorship is assigned to individuals according to the magnitude of their scientific contribution to the study. Research suggests that this principle is generally followed in psychology (Over and Smallman 1973). Whether a person is being paid or receiving class credit for his work is irrelevant in considering authorship: scientific contribution is the key consideration. Factors such as power or the sex of the authors should not be considered (e.g., Wilkie and Allen 1975). Whether an investigator is a world-renowned scientist or an undergraduate research assistant is not a consideration except insofar as his role has influenced his contribution to the study. Last, the actual number of hours spent on the study is secondary to the scientific contribution. Authorship should not be given simply as a reward for hard work. The key question, then, is: What constitutes a "scientific contribution"?

Scientific contributions determine the content, scope, and interpretation of a study. The two specific activities that normally merit authorship credit are conceptualizing and designing the study and writing the report. Each of these will determine the content and character of the study and the publication. In contrast, a variety of necessary activities could be completed by many people and do not in themselves determine the scientific character of the publication. In general, such a job does not justify publication credit. For example, typing the report is a nonscientific activity and will not affect the scientific merit of the study. Other activities that usually do not deserve authorship credit are computer programming for data analysis, routine running of subjects, and clerical work.

A survey of psychologists by Spiegel and Keith-Spiegel (1970) helps delineate the specific activities that scientists think deserve authorship credit. They sent questionnaires to a group of psychologists describing studies where people worked together and asked how authorship should be determined in each case. The group surveyed was large and diverse, with 746 respondents. Most were productive scientists, with an average of twenty-three publications each (about half of these coauthored with others). Therefore they had considerable publication experience and had frequently been involved with others in determining authorship. Those surveyed believed that designing the research procedures and writing the final report were the scientific tasks that should mainly determine authorship. The researchers believed that the data collection and analysis were usually less important. However, it should be noted that in fields like anthropology the actual data collection may not be routine and may greatly influence the study's results.

It should be clear that to deserve authorship credit an individual's contribution should be both substantial and scientific. A small contribution—for example, recommending a minor modification in research design—does not deserve authorship credit; it should go to those few persons who contribute most heavily to the planning, interpretation, and writing of a study. Authorship should only go to those who are actively involved and should not be given to others out of gratitude, deference, or friendship, as when a laboratory director requires that all publications from "his" laboratory give him an authorship, when someone with a large data set trades access to data for publication credit, or when a professor makes minor revisions to a student's paper and adds his own name as second author. There are cases where such authorships are justified. For instance, a professor may contribute substantial changes that make a student's paper publishable. When someone uses an existing data set, the person who collected it may deserve authorship because he essentially designed the study by the way the data were originally collected. Occasionally someone wants to give joint authorship to a well-known researcher because he thinks it will improve his chances for publication. Anyone whose contributions have not been significant should reject such an offer of authorship.

In student-faculty interactions, the faculty member normally determines authorship, and this may work to the disadvantage of the student. On the other hand, there is one situation where the student often takes advantage of the professor. Sometimes a master's thesis or Ph.D. dissertation is the advisor's idea or stems directly from his research, with the student perhaps being paid to run it. With planning sessions and revisions of writings, the professor may have contributed more ideas and work to the design and write-up than the student. Yet it is common for the student to later publish the dissertation as sole author. One reason for this is that dissertations are often supposed to be independent contributions. If a professor later shares authorship it suggests this criterion may not have been met. Some disciplines have written or unwritten rules that professors cannot be authors on student theses. Professors should advise students as part of their jobs; but there are many cases where students should be told that if extensive help is to be rendered with the idea or the writing, the faculty member should be given second or even first authorship.

Some specific examples of contributions may help clarify questionable cases. A person who only analyzes the results of a study usually does not deserve authorship unless the analysis represents an important contribution to understanding or rethinking the study. A person who only collects data usually should not be an author, whereas a person whose contribution also includes planning the study or writing it up will deserve authorship. Paid research personnel are entitled to the same authorship privileges as others if they make similar contributions. If a research assistant works independently in running a study and contributes to explaining the results or writing the report he probably deserves junior authorship. If this is in doubt, the assistant can be encouraged to contribute to the write-up to merit an authorship. An investigator who helps plan a study but then does not follow through with aid in running and interpreting it should usually not receive authorship unless his planning contributions were major. If a colleague or acquaintance mentions an idea for research and you decide to implement it, he should be asked if he would like to collaborate on the study (and, if so, be given authorship credit). If not, he will usually deserve only footnote credit unless his original contribution was extensive.

It is difficult to collaborate equally on research and to allot authorship fairly. Occasionally disputes arise over authorship. The best way to avoid such disputes is to make an agreement about authorship and division of labor during the planning stage. Such an agreement should include the proposed contributions of various individuals and also a rough timetable. Sometimes one contributor procrastinates, and a set of deadlines can help clarify this. If a person's procrastination becomes so flagrant that others must take on his duties, the person should usually lose authorship. Scientists have an obligation to their colleagues to meet agreed-upon deadlines. Where they totally fail to meet their obligation, they can be given footnote credit if they have made some contribution or junior authorship if they made a substantial contribution but failed to complete the work.

Guidelines for arranging authorship must always be tempered by individual circumstances. Each case contains a unique combination of contributions. Often each individual contributes something to each stage of the research. Each remembers his own input clearly but seldom sees all the work the other contributors do. When there are disagreements about relative contributions, those with less power or status are at a disadvantage. However, Spiegel and Keith-Spiegel (1970) showed that people with many publications were actually more generous in how they said they would assign publication credit than those with little publishing experience. Similar results have been found for Nobel laureates (Zuckerman 1968) and distinguished psychologists (Over and Smallman 1973).

Many disagreements have arisen in determining publishing credits.

In one case a graduate student carried out a study designed by a professor, requiring continual help, encouragement, and supervision. He then analyzed it and added an important subanalysis that turned out to be the most interesting result. The student felt he deserved first authorship because of all the work he had invested. But this was contrary to the general guideline that scientific contribution takes precedence over the amount of work. In another case a paper was tentatively accepted by a leading journal pending revision. The first author was a professor, the others were graduate students. Several times the first author promised to do the revisions by a certain deadline. He did not, and after more than a year the paper was dropped by the journal. One of the other authors then undertook the revisions but had to resubmit the paper. Perhaps a more typical case is one author's delay while data are being analyzed and written up. He may agree to do a draft by a certain date then miss the deadline because of other demands on his time. In one case we know, the second author refused to accept the draft done by the first author yet did not revise it to his own liking for more than a year.

Guidelines for determining authorship could avoid many of these problems, as could preagreements defining each person's responsibilities and the order of authorship. Where individuals do not live up to the agreement, authorship should be adjusted accordingly. If disputes about authorship or research obligations cannot be settled among the authors themselves, a third party (e.g., a department head) can often mediate or decide the issue. One other type of dispute is a disagreement about the contents of the report. Where there are substantial disagreements, the senior author's version should be published and footnotes can indicate matters on which other authors disagree. For further discussion on authorship, see De Solla Price (1964) and Spiegel and Keith-Spiegel (1970).

MULTIPLE PUBLICATIONS

Several questions besides authorship arise in the ethics of publishing: for example, Can the same article be published in several journals? In general, multiple submissions or multiple publications of the same article are considered undesirable because of the shortage of journal space and editorial time (e.g., De Solla Price 1964). Also, copyright infringements are likely when the same material is published in more than one place. To disseminate information to different audiences, separate papers can be prepared that properly cite related articles (Gardner 1971). Usually an article should be submitted to only one journal at a time for consideration for publication so that editors' and consultants' time will not be wasted. If a journal rejects a manuscript, the authors can then submit it to another journal. A few journals allow multiple submissions, but some disciplines such as sociology have explicitly banned it (Peters 1976). If a person does simultaneously submit a manuscript to more than one journal or book publisher, he is obligated to mention this to the editor in his transmittal letter so that the editor can decide whether he wishes to review an article that is also under consideration elsewhere.

The major reason authors want to submit papers to more than one journal is the low odds of a paper's being accepted and the long wait for a decision. For instance, McCartney (1976) reported that the median rejection rate for thirty-nine sociology-related journals was eighty-two percent. It often takes up to a year to reach a decision on a paper and another year or more before it is published. Much of the responsibility for this lies with slow or overworked journal editors and reviewers (e.g., Rodman 1970). Peters (1976) has argued that forbidding multiple submissions protects major journals from competition and gives authors less power (see also Schuessler 1976).

With large studies a question often arises about how many articles are needed to communicate the results. Obviously, one should not write several articles where the information can be conveyed in one, but this sometimes occurs because authors want to list more publications on their vita. It is unethical to fragment a study into several articles simply to gain additional authorships. Not only does such a practice waste journal space, it makes it harder for readers to understand the entire study. In fact it is often desirable to combine several programmatic studies into one article so readers can better grasp the findings from the entire sequence. On the other hand, sometimes it is necessary to write several distinct articles about a study, especially if it is very large, either because there is too much information for one manuscript or because several distinct issues were addressed by a single collection of data. When more than one article is published from a single data set, each article should specify this and cite the other articles so that readers can find the related information and so that editors are aware that a larger study is involved.

169

Grants

Social science research is frequently supported by grants or contracts from public and private agencies. Agencies such as the National Science Foundation and National Institute of Mental Health base some grants on the quality of unsolicited research proposals. At other times agencies request proposals on specific topics and support the research that best fulfills their goals. A grant gives the researcher more freedom than a contract, which usually means that a researcher is hired to collect specific types of information. A variety of ethical issues arise in the area of grants and contracts because the scientist may misuse the funds or the granting agency may make improper demands on the scientist.

RELATION TO THE GRANTING AGENCY

Researchers must face the important question: Do I believe this research is worthwhile or am I doing it only to get the funds? An investigator should not apply for funding if the goals of the research or the agency seem ethically questionable. For example, if a scientist were asked to do research on how to manipulate public opinion for political gains, he might well refuse. Some investigators refuse to accept funds from certain agencies such as the Department of Defense because they disagree with their goals or philosophy. Others will accept funds if they believe their findings will not be used for bad purposes. Each investigator must decide whether he should accept funds from a particular agency and whether the research is justified. For some institutions, such as the National Science Foundation, a major purpose is the support of scientific research, but other agencies are "mission oriented," and the researcher should be aware of how his research may be used to further the goals of that agency (e.g., national defense; see also chapter 13). Regardless of the source of funds, the scientist should not accept money unless he believes his findings will be used constructively.

Investigators should also ask whether a sponsoring agency will try to influence the outcome of the research. For example, it may try to bias the results by the sample it chooses, as in the American Medical Association survey of the elderly discussed in chapter 9. In such a case the scientist should refuse to conduct the research, for he should not subvert the scientific method to gain results that will please the sponsoring agency. An agency may also want to control the dissemination of the findings or even forbid publication of

170

unwelcome results. Except for reasons of national security, such restrictions on the free dissemination of information should be strongly opposed on ethical grounds. Another possible exception to the ideal of open dissemination of research is when a company sponsors industrial research to develop a new product or service and open publication might help competitors. The general principle is that whenever a funding agency seems likely to undercut the objectivity of the published findings, its support should be rejected. Similarly, restrictions on the open dissemination of research findings should be avoided unless there are very sound justifications for them.

Another potential problem arises when the sponsoring institution, especially a private agency, uses an outside researcher to collect sensitive information, possibly about the organization itself, then tries to force the researcher to violate the anonymity of the individuals or groups studied. The officials may decide they want to know who answered questions in certain ways. For example, a company might hire an industrial psychologist to study how to improve employee morale, then pressure him to disclose which employees favor a union. Where the sponsor's interests conflict with guarantees made to respondents about anonymity, the researcher must strongly support the respondents' rights. If proper safeguards to privacy were established as discussed in chapter 4, the researcher can destroy master lists or go to court to maintain secrecy. Further discussion of the relationship between sponsors and researchers is contained in Orlans (1967).

GRANT PROPOSALS

Grant proposals are supposed to be treated as confidential by agencies and reviewers. But reviewers sometimes take advantage of seeing material before it is published. For instance, they may leak information to colleagues (e.g., Davis 1969), steal ideas, or delay or reject the grant to hinder competitors. Needless to say, these behaviors are unethical.

On the other hand, because of the time invested in writing most grant proposals and the delay before acceptance, researchers use several procedures that can be ethically questionable. The scientist may do the proposed research before the grant is received, then use the grant money for related studies or for another project. Or a grant proposal may be submitted for research that is already completed. In these situations the researcher does the work specified

171

in the proposal, but the funding requested is not really necessary (Borek 1976).

An even more questionable practice is to apply for funds for a proposed project, then use the funds for entirely different work. It is sometimes permissible to use a small amount of resources from a funded project for related projects, but contract terms often require permission for any reallocation of money. In a large research center it sometimes becomes unclear what money is used for what project, since the same people may work on more than one thing. An extreme example of this is given by Riesman and Watson (1964), who cite a case where the funded project was not ready to go when the grant money was received and much of the money was spent on a related project.

Frequently a scientist will propose a series of studies that are funded as a package; but after conducting the first study, it may become clear that the other proposed studies are no longer desirable. The results of the first study may suggest alternate studies that will be more fruitful. When the modifications change the original studies only by improving their ability to meet the stated objectives, it is usually permissible to conduct the improved studies without informing the sponsoring agency. But when the studies conducted will be entirely different from those originally proposed, it usually is desirable, and sometimes is required, to notify the granting agency. Finally, it should be obvious that the researcher has an obligation to complete research by the date specified in the grant or contract.

For further reading, see Rodman 1970 and Spiegel and Keith-Spiegel 1970.

11 The Subject Pool

Psychological research uses students as subjects so frequently that the discipline of psychology is often characterized as "the science of the college sophomore." Reviews of published research have shown that an overwhelming majority of subjects in psychology experiments are college students (Carlson 1971; Menges 1973; Schultz 1969), a large proportion of them recruited through an organized "subject pool." In other disciplines also, volunteers are often solicited in classes, or class time is used to conduct studies. At times student participation in experiments is required, at other times it is rewarded by extra credit, and sometimes it is completely voluntary. Since a large number of college subjects are recruited in nonvoluntary or semivoluntary ways, it is not surprising that there has been much debate about the ethics of classroom experiments and research participation requirements. For further discussion see also the APA Code of Ethics, 1973.

The Case in Favor of Subject Pools
The subject pool benefits the researcher because he can acquire subjects without monetary expense. Therefore, the hours donated by students permit important research that often could not be completed otherwise. In addition to the benefit society derives from the research, subjects themselves benefit in several ways from an increased level of research at their school. Research is supported because it advances knowledge. This knowledge is eventually conveyed to students, improving their education. Thus students who act as subjects benefit from research knowledge much more directly than does the general public because they themselves are studying the subject (King 1970). Although beginning students are rarely knowledgeable enough to contribute ideas to the discipline, they can contribute to increasing knowledge by participating in research, and they in turn benefit from the participation of past students.

The most reasonable justification for maintaining subject pools is

173

that participating in research is educational and therefore directly benefits the students. Seeing experiments in progress in the laboratory gives the student firsthand experience of what research is like. In addition, many universities require that students be further educated after the study by an explanation of the research and its purposes. Although they may not remember the exact design or the variables studied in complex research (Menges 1970), they do learn about the process in general. Although only a few students believe participation is a great experience, most do feel afterward that it was a worthwhile learning experience (Davis and Fernald 1975; Gustav 1962; King 1970). In many ways the participation requirement is like the laboratory section in traditional science courses, with the addition that the subject is contributing to knowledge, not just going through a learning exercise. Since students are usually allowed the alternative of doing something of equal educational value, their freedom is not violated.

The Case against Subject Pools

Educational benefits to students are often used to rationalize the subject pool or "sell it" to higher university authorities. The subject pool is often of benefit only to the people doing research. Frequently the studies are boring and have little educational value for participants, and in practice researchers may do little to make research participation profitable for subjects (Menges 1970). The requirement to educate students about the study often is omitted or fulfilled only by giving subjects a dry mimeographed sheet explaining the study. Where students are given an alternative to participation to insure their freedom, it is sometimes something onerous and distasteful like writing a ten-page paper. Participation is least voluntary when bonus points are given for volunteering and the competition for grades is keen, or when participation is required in a course that is not elective. It is difficult to insure voluntary participation when students participate as part of a course (Arellano-Galdames 1973) and the professor thus has power over them.

A substantial number of students resent the participation requirement and may therefore treat studies as a game or consider the experimenter an opponent. If they spend their time trying to "psych out" or understand the experiment, they may not react naturally. Even worse, student subjects may deliberately behave abnormally or give dishonest responses to sabotage the experiment (e.g., Brower

174

1948; Cox and Sipprelle 1971). This destroys the value of the research.

Resolution

Confronted with arguments over the ethical nature of subject pools and other uses of students for research, professors may wonder whether they should support such practices in their departments. Young researchers, students, and professors may question whether they personally should use subject pool participants. It should be noted that subject pools are only one of the ways students are involved as semivolunteer participants in experiments. For example, many teachers in social science and education courses use class time to collect questionnaire data from their students, and professors may require that students themselves collect data as part of a class project. Like the subject pool, these situations may pose a conflict between educational and research goals.

The reader may have noticed that most of the arguments against the use of students as subjects are based on the thoughtless and uneducational way subject pools are administered and do not reflect on all research activities involving students. Participants from a subject pool can be ethically used in research if certain safeguards are instituted and the abuses mentioned above are avoided. Guidelines are therefore needed. However, no matter what formal procedures are instituted, the ethicality of the subject pool will depend on how student subjects are treated in practice. Even when the formal guidelines governing the subject pool seem satisfactory, researchers using student subjects may violate ethical norms by mistreating subjects or ignoring their obligation to educate student subjects.

To guarantee students' freedom to choose whether to participate, several safeguards should be instituted. First, it should be publicized that any student who objects to participation for whatever reason can meet the requirement in some other way. The alternative requirement is justifiable since research participation is itself an educational experience. The alternative to participation might be reading a paperback book, watching movies of laboratory experiments (see Davis and Fernald 1975), making informal field observations of behavior, and so forth. It must be educational and must not be more onerous or time-consuming than the participation hours. Where credit is given as bonus points and participation in research is not required, the grade curve should be computed without considering bonus points, and bonus points should also be available for other educational experiences. These policies avoid imposing a penalty on nonvolunteers.

Another procedure that helps guarantee freedom of participation is allowing subjects to withdraw from, or refuse to participate in, particular experiments. This freedom should be clearly announced when the participation requirement is explained in class, and it is ideal to remind subjects of it when they arrive for a study. Whenever a specific study entails risk or discomfort, subjects will usually be asked to sign an informed consent form. The subjects' freedom to refuse participation or to withdraw from the study should be clearly stated in the form. Where no informed consent form is used in nonrisky studies, student subjects should be reminded at the beginning of the investigation that they may withdraw at any time. Students may also be given an advance option of declining to participate in studies that involve deception or stress, and thus they will not be called for research that includes these elements.

A third safeguard for insuring that student's participation is voluntary is to state the participation requirement in the course description in the university catalog. This will allow students who do not care to participate in research to choose other courses if this one is not mandatory. Although students cannot choose to skip the laboratory sessions in most science courses, more freedom is desirable when students meet a laboratory requirement by serving in research because their participation serves both an educational *and* a research role.

A second set of guidelines for proper use of subject pools is directed at making the experience truly beneficial for student subjects. Every effort should be made to make participation educational (see, for example, Davis and Fernald 1975). Many subjects come to experiments curious and sometimes even eager about this new experience (Gustav 1952), and their curiosity should be encouraged by an enthusiastic explanation of the study afterward. In this way we may also guarantee that future generations of the college-educated public will have positive attitudes toward social science research. An experimenter should explain the research clearly, in a friendly way. This educational obligation is especially great when the subject has probably learned little experientially, as when he has completed a lengthy general attitude survey or been in a control condition. A written description of the study may often be a useful supplement to the oral explanation.

The number of research hours required should not be too high. The educational value of being a subject is greatest during the first several hours of participation, and the amount to be learned probably drops considerably after three or four experiments. Only when experimental participation is truly educational is the requirement comparable to laboratory sections in the physical and natural sciences.

Several additional precautions may help insure benefit to students. Some channel for complaints should be open through the department chairman, the research committee, or the course instructor. Students who feel they have been mistreated or had their rights violated, or who have not profitably learned about the research from the investigator, can then issue a complaint. This feedback may motivate investigators to treat student subjects more responsibly. If many complaints occur about a particular experimenter, his right to run subjects could be withdrawn. It is also important that proposed studies be approved by an ethics review committee to insure institutional protection. It is desirable to include students on this committee since they are the most frequently used subjects.

Conclusion

A subject pool can be ethical and educational, the equivalent of a laboratory section. If a scientist supports the use of a student subject pool or employs participants from the pool, he should insure that: participation in a particular study is voluntary, not mandatory; participation is educational; subjects are treated with respect; studies using the subject pool are approved by an ethics review committee. Since most student research (e.g., Ph.D. dissertations, masters theses, undergraduate honors theses) is unfunded, laboratory studies by students usually rely on the subject pool. In this way the subject pool serves a dual educational function—for both participants and researchers.

For further reading, see Davis and Fernald 1975; Gustav 1962; King 1970.

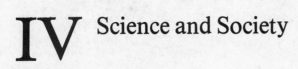

IV Science and Society

12 Science and Values: The Input of Values to Scientific Work

> Scientific objectivity does not mean the abandonment of moral conviction and belief, but rather the willingness to control one's convictions and beliefs in order to review evidence in a dispassionate way.
>
> Arnold Green (1972)

> Even with utmost care, a biased view cannot be avoided. It can only be counteracted insofar as it is made explicit and confronted with analyses based on alternative perspectives.
>
> Donald Warwick and Herbert Kelman (1973)

The traditional image of science has been one of dedicated individuals seeking objective truth (Lundberg 1947). When the term "value free" was applied to the social sciences it reflected the belief that scientists' values and personal opinions should not influence their scientific inquiry. The "ivory tower" scientist was supposed to stand apart from the world, examining phenomena in a dispassionate and objective way. The only values that were supposed to influence research were the scientific values placed on truth, objective methodology, and the open distribution of knowledge. Recently the value-free conception of science has been challenged and important issues have been raised about the appropriate relationship between scientific work, objectivity, the role of science in society, and scientists' values and beliefs. In this chapter, we will discuss the influence of values on scientific work and the extent to which science can validly be called "objective."

According to the traditional view of science, the researcher's task was to uncover objective knowledge about the material world. The term "objective" has several interrelated meanings; it is used here to mean a description of "facts" not distorted by personal beliefs. The term "facts" assumes a particular belief about the nature of reality. Largely on the basis of philosophies such as logical positivism, scientists assumed that there was a material world "out there" with properties independent of the observers, and that the true properties of this world could be discovered by careful empirical research. Each study was seen as uncovering new knowledge about the world so that scientists were slowly coming to know the world as it really was. Scientists were to pursue science for its own sake because knowledge itself was thought to be a worthwhile goal. Indeed, the logical positivists went even further in claiming that value statements were meaningless and therefore could not possibly be a part of science.

The extreme traditional view prescribed complete detachment between personal values and scientific research. The scientist was to select and approach problems from the viewpoint of science, indifferent to the potential uses of his findings. Social scientists were viewed as impartial observers, recording and analyzing human behavior and creating theories to explain it. Scientific findings were to be communicated dispassionately to policymakers and the public, who could then use or ignore them as they saw fit. In conducting research and presenting findings, the cardinal rule was, "Thou shalt not commit a value judgment" (Gouldner 1963, p. 35). It was permissible for scientists to express opinions and values as individual citizens, but as scientists they were professional producers of knowledge who were only to present an accurate rendering of the facts for society to consider.

The principle of a value-free science probably arose historically in order to insulate scientists from demands by powerful persons and groups to fit their theories and findings to predetermined conclusions. However, the purpose of value-free science went beyond merely protecting scientists from outside interference. Scientific methods are designed to gather knowledge and cannot be used to decide which values a society should hold and what goals it should work for. The value-free image of science thus attempted to limit scientific endeavors to the realm for which they were best suited. Take the case of a researcher studying the influence of television violence. The scientific method can help find out whether viewers are more likely to be aggressive after watching media violence, but it cannot determine whether violence is morally good or bad. Thus, the value-free view attempted to keep scientists within the limits of their true expertise.

Decline of the Value-Free Image

Despite the advantages inherent in the value-free image of science, developments within society and within the discipline of philosophy have now made us aware that a completely value-free science is impossible for a variety of reasons and that the traditional conception of science as discovering principles about the "real" world was rather naive (Kelman 1968). Although one determinant of scientific research is the scientist's desire to search for knowledge, there are certainly many other influences as well. First, research topics and methodology are usually decided by factors such as personal interests and values, the desire to prove a point, the availability of money, the influence of teachers and colleagues, and other concerns outside of science. For instance, governments may influence what research is done by offering millions of dollars to support the study of specific topics: for example, child abuse, ghetto life, or reading skills in the schools. Also, what phenomena are studied is often determined by the social scientists' values. For instance, the sociologist who is personally afraid of the possibility of nuclear holocaust might investigate peaceful methods of conflict resolution, and the anthropologist who values equality may study ghetto life. However, even "lesser" motivations can influence scientific work. Nobel laureate James Watson, the codiscoverer of DNA, described in *The Double Helix* (Watson 1968) some of the human frailties and other personal factors involved in scientific discovery, including jealousy, competition, secrecy, and the luck of being in the right place at the right time.

Another factor that makes complete value neutrality impossible is that problems are defined and questions asked in terms of the experiences and beliefs of the social scientist. For example, a sociologist's desire to study poverty often reflects his values; and, beyond that, whether he believes poverty is caused by people's laziness or by inequality of opportunity will influence the types of data he collects. One scientist might study why the poor are low in achievement motivation, whereas another might investigate the entrenched power structure that prevents them from receiving a quality education. How the particular scientist conceptualizes the problem will reflect his values, political orientation, and general beliefs about human nature. Though scientists studying poverty may all be concerned with improving the lot of the poor, their initial beliefs and value orientations will influence the questions they ask and therefore partly determine the answers they uncover.

Riegel (1972) has argued that cultural and philosophical assumptions influence the development of science. Similarly, the scientist's discipline will also channel his thinking. Psychologists ordinarily are trained to think in terms of person-centered principles (Caplan and Nelson 1973) and are therefore likely to examine poverty from an individual perspective that focuses on attitudes, learning, and motivation. Sociologists are more likely to focus on the social system. For example, in studying upward mobility, a psychologist might focus on need for achievement, while a sociologist might focus on class structure. Scientists within both disciplines often ignore approaches that lie outside their own area's traditional type of analysis.

Even rigorous methodology cannot guarantee an absence of bias and value influence. Once the data have been collected, interpretations can vary widely, depending on scientists' theoretical commitments and other beliefs. Arthur Koestler (1971), in *The Case of the Midwife Toad*, traces how prior theoretical commitments determined scientists' acceptance of certain empirical findings in the debate between Lamarckism and Darwinism. The race and IQ debate further illustrates how value commitments and beliefs may influence the way different scientists interpret their findings and may determine which data they focus on. Thus we can see that the scientific process, from choice of problems to interpreting findings, is filtered through the values and personal idiosyncrasies of individual scientists as well as being influenced by societal factors.

The value-free image of science has also declined because the idea of a completely "objective" science has been challenged by philosophers of science such as Kuhn (1970) and Polanyi (1964). We cannot know about the external world in any absolute sense; we can only have experiences that come to us through our senses and are organized in differing ways. There can therefore be many different "realities," depending on the cognitive categories and perceptions of the knower. There is no one correct way to conceive of the universe, but a multitude of ways, depending on one's culture and thinking. As Kuhn pointed out in *The Structure of Scientific Revolutions*, even within science there have been many "paradigms" that have differed in the assumptions made, in the questions asked, in the phenomena observed, and in the ways the data were interpreted.

Those operating within different paradigms often believe different events are most important and conceive of the same events in different ways. Science is now seen not as the steady, ever-increasing objective growth of knowledge in a single direction, but as a series of discrete advances in somewhat different directions. In the social sciences there are many different theoretical orientations that lead social scientists to describe events in diverse ways. Although Kuhn's original position has been criticized by historians and philosophers of science (Wade 1977), its impact upon our view of science has been dramatic. Probably the major contribution of Kuhn's influential book is the increased realization that logic and empirical findings are not the sole arbiters determining which scientific theories are accepted as correct.

Value issues also arise in the social sciences because scientific theories and findings often represent powerful social forces in themselves. In physics, Heisenberg stated that observing a subatomic particle may alter it. In the social sciences, observations and discussions of findings are even more likely to influence what is being observed, since people, the subjects of study, can understand and be influenced by the scientist's observations. In anthropology, for example, the presence of the anthropologist may change the culture studied, sometimes dramatically. In psychology and sociology too, the scientist's interest in a phenomenon may alter it.

The publication of scientific theories and findings may affect behavior. Rainwater and Pittmann (see chapter 6) knew that their description of a socially deviant community in Saint Louis could have large repercussions for the group. Kinsey's report of sexual behavior may have influenced the sexual practices of later generations. Scientific predictions about future societal trends may end up being wrong because the prediction itself affects the direction society chooses. And nowadays powerful government policymakers may enact laws and policies based upon the findings and predictions of social scientists, thus altering the course of society. As Melvin Webber (1973, p. 203), an urbanologist, wrote, the scientist "cannot escape the fact that his facts are instruments of change. To play the role of scientist in the urban field is also to play the role of intervener, however indirect and modest the interventions." Furthermore, Webber cautioned us that "Seemingly straightforward facts about a society's things and events are seldom, if ever, neutral. They inevitably intervene into the workings of the systems they describe.

185

The information supplier—whatever his methods and motives—is therefore inevitably immersed in politics. The kinds of facts he selects to report, the way he presents them, the groups they are distributed to, and the inferences he invites will each work to shape the outcomes and subsequent facts." Hence, even when social scientists present findings in a dispassionate and unprejudiced way, their "objective" scientific descriptions may alter the very thing being studied. In addition, the "scientific issues" addressed by the social sciences are often composed to a large extent of extrascientific factors (Nelson 1975) such as cultural definitions, assumptions, and political ideology.

In conclusion, a host of developments and ideas have combined to put to rest the tenet that social science can be value free. The argument has even been made that science can both study and shape societal values (e.g., Sperry 1977). Nevitt Sanford (1965) anticipated the current view when he stated that the problem with value neutrality is that in the last analysis it is impossible. The danger with supporting supposed value neutrality is that our values and their influence will go unrecognized and thus prevent us from assessing their possible biasing influence.

The Activist Conception of Science

A group of "radical" social scientists has attacked the argument for value neutrality even more harshly and argued for a new conception of the place of science in human affairs (Ehrlich 1969; Gouldner 1963; Gray 1968; Mills 1959). As David Gray wrote in his Ode to Behavioral Science:

> Ethically neutral, value free—
> tweedle dum, tweedle dee.
> Nazi S.S., Schweitzer humane—
> value free, all the same.

So-called neutral research, they maintain, is usually research that supports the status quo. Values and personal factors exert a heavy influence on all research, but science often is made to appear completely objective, with the unexamined values and assumptions hidden because most people in the society share them. The radicals maintain that it is impossible to do value-free research and that therefore all research is political. The radical view appears to have two major messages. One is that social scientists must study social

186

issues and social problems. In principle this argument has been both advanced and practiced by some major social scientists. For example, C. Wright Mills (1959), a sociologist, and Kurt Lewin (1948), a social psychologist, have both argued that scientists have a responsibility to try to understand and correct human problems. The second message of the "radical" group is that much "value-neutral" research has actually been partisan research in disguise. Much of the controversy over heritability of intelligence and race demonstrates that what seem to be objective scientific discussions may be heavily biased by assumptions, beliefs, and choice of data. In the IQ debate, some scientists have argued that blacks are probably innately inferior in intelligence. Many others have attacked this position and argued that the poor showing of many blacks on IQ tests is due either to the cultural bias of the tests or to the impoverished environment of poor blacks. The data have been inadequate to clearly support either side. In such a case, prior beliefs and value commitments frequently have determined the sides people have taken in the debate and the evidence they have marshaled in their favor. Hirsch (1971) and others have outlined the impact of nonscientific beliefs and values in this "scientific" debate.

The activist social scientist also believes in becoming involved in translating findings into policy, whereas the traditionalist will usually let the "facts speak for themselves." The dichotomy of two types of scientists is of course an oversimplification—there is a whole range of opinion about the appropriate degree of scientific involvement in social problems. At one end of the continuum are those who choose research problems because of their theoretical interest, with no regard for their potential application. At the other extreme are scientists who believe that research should be done only if it directly attacks social problems. In the middle are those who recognize that some scientists will usually attack problems without immediate applied impact while others will do research on applied social problems. However, many activist social scientists such as C. Wright Mills recognize that the scientist's major contribution to the solution of social problems is nevertheless scientific, not political. Mills (1959, p. 192) wrote:

> The role of reason I have been outlining neither means nor requires that one hit the pavement, take the next plane to the scene of the current crisis, run for Congress, buy a newspaper plant, go among the poor, set up a soap box. Such actions are

often admirable, and I can readily imagine occasions when I should personally find it impossible not to want to do them myself. But for the social scientist to take them to be his normal activities is merely to abdicate his role, and to display by his action a disbelief in the promise of social science and in the role of reason in human affairs. This role requires only that the social scientist get on with the work of social science.

The Emergent View of Science

It is now recognized that all knowledge is to a certain extent subjective. However, the methodology of science is primarily designed to minimize the possibility of errors or bias in collecting and interpreting information. No matter what area the scientist is working in, his unique contribution is the empirical perspective; data and scientific theory, not dogma or prior beliefs, are the arbiters of truth. The scientist seeks knowledge through methodologies and techniques that guard against personal bias, then accurately reports the findings and their limitations. Other scientists, often with different personal viewpoints, will seek to replicate, challenge, or extend his findings. In combination, different studies on the same topic by a number of researchers may accumulate to produce knowledge less biased than that found in a single study. Take, for example, a sociologist who believes strict gun controls would reduce homicide rates. Despite his beliefs, he realizes that his opinion may be incorrect and that gun control may have little effect on crime. He might examine homicide rates before and after gun controls were instituted in several places. Another scientist who doubts the desirability of gun control might attempt to replicate the findings of the first researcher. The second scientist might examine time-series intervention data from other areas or might use a different methodology, for example, correlating the enforcement of gun control laws in different locales with the homicide rates of those areas.

Whether or not a scientist initially believes in the efficacy of gun control, he should collect evidence objectively and then critically analyze it from all perspectives. The scientist does not set out to merely collect anecdotal evidence or evidence from preselected areas that will prove his point. If the scientist uncovers evidence that gun control laws are not effective, he analyzes this evidence objectively and publishes the results. Because of his allegiance to truth, he presents his evidence and methodology carefully, in a standard way.

Whereas a journalist may be intent on swaying readers with rhetoric and emotional appeals, the scientist must accurately present the facts he has gathered. He may then describe his conclusions and suggestions based on the research, but he should also present the limitations of his research, thereby helping others to improve upon the findings and to know where potential errors lie. Others may thereafter challenge his interpretations of the findings, attempt to replicate the study, and conduct more sophisticated work on the question. Although a scientist often must make educated guesses based upon inference and incomplete information, these limits should be clearly set forth when the conclusions are expressed. C. Wright Mills said it well: "In such a world as ours, to practice social science is, first of all, to practice the politics of truth" (1959, p. 178).

Part of the confusion over the value-free nature of science arises because of a failure to differentiate scientists' motives and methods. The traditional view of science as value free is critical for the method of science. In carrying out investigations, *objective methods* must be used and various interpretations of the data considered in an open-minded way (e.g., Becker 1967). Social scientists use methods like "blind" recording, reliability checks, and statistics to prevent themselves from letting their prior opinions mislead them. An example of objectivity is a scientist who surveys political attitudes for a forthcoming election. Although she knows how she plans to vote, she formulates objective methods of polling (e.g., random sampling and neutrally phrased questions) that minimize the possibility that her own preferences will influence the outcome of the poll.

In contrast, the newer value-laden view of science is helpful in understanding the *motivations of scientists* and the possible uses of the scientific method. Personal interests will help stimulate the social scientist to devote attention to a particular problem. We can thus see that there is no inherent contradiction between humanistic motivations to attack certain problems and the use of objective empirical methods. The distinctive feature of science is that, although the scientist may begin with a personal opinion, he pursues the truth wherever it may lead (Sanford 1965).

Yet scientific objectivity cannot be based completely on rigorous methodology. Although methodology is helpful, the questions asked will still be affected by the scientists' cultures, interests, and values

189

—and an objective methodology cannot overcome this initial bias. Ultimately science, like other approaches to knowledge, will be influenced by scientists' cultural and personal characteristics. As MacRae (1976) suggests, the social sciences are not simply empirical. Besides empiricism, logical thought, theories, and values also aid in the understanding of behavioral and social phenomena. Once factual questions are formed, the scientific method has many built-in safeguards to increase the accuracy of the answers uncovered and to minimize the influence of personal biases or expectations. Nevertheless, the scientist should constantly consider his own biases and how these may be influencing the research. He should be on guard not to let subjective or ideological factors determine the research outcome in favor of certain conclusions.

In addition to an empirical approach to knowledge, scientists as a group work against bias by bringing to problems an ability to consider alternative perspectives. In fact, one recommended way to design research is to use strong inference so that any result is informative and the scientist is testing several approaches at once rather than trying to support a "pet" theory (Platt 1964). Seidman (1977) recommends that the scientist bring at least two radically distinct conceptualizations to bear on a problem when formulating a research strategy. The scientist should also be revisionary in that he is willing to apply new conceptions to problems. For example, for years society has defined violence in purely physical terms. This has helped to brand those who commit violence as immoral deviants and to excuse societal structures that probably cause much of the physical violence. Realizing this intimate connection between the society's structure and violence, Galtung (1975) has formulated the concept of structural violence—that violence can occur indirectly when societal structures benefit certain persons and simultaneously harm other individuals or classes of individuals. He hypothesizes that physical violence usually cannot be abolished unless structural violence is also eradicated. Thus Galtung has adopted neither the view of the "establishment" that only personal violence is bad nor the radical view that violence is usually good if used against societal structure, but has sought to redefine the issues in a new way that synthesizes both viewpoints.

Another example of a scientist defining issues in a new way is the extensive theory of Sigmund Freud. Freud conceptualized problems in revolutionary ways and brought a questioning attitude to phenomena that had been taken for granted. Scientists following Freud then applied more rigorous empirical methods to test and refine his ideas. Herbert Kelman (1968, p. 4) ably summed up the relation between science and values, and the delicate balance between humanistic concern and objectivity:

> Social research, in my view, must be based on the recognition that neither the social researcher as a person, nor the process of social research can be entirely value-free.... What is necessary, however, for the enhancement of scientific objectivity, is that the investigator deliberately takes his values, attitudes and expectations into account and systematically analyze their effects on the definition of the research problem, the observations obtained, and the interpretations placed on these observations. We can never eliminate the effects of values and subjective factors, but we must push against the limits to scientific objectivity that inevitably govern our efforts. In my view, the tension between the investigator's values and a recalcitrant reality world, and the constant—though never wholly successful—effort to disentangle the two, are central and constructive features of the scientific study of man.

Applied versus Basic Science

From the debate over the form science should take, there emerges a synthesis that recognizes legitimate differences in the ways individuals may "do science," while still maintaining that certain fundamentals are common to all scientific activity. Developing accurate knowledge and valid theories is the primary expertise of scientists, but they may individually choose to use this skill in different ways. Some scientists may do purely theoretical work unrelated in any immediate way to world problems. Evaluation of *basic* research, as it is called, should be based upon the theoretical advances it produces, not on its immediate effects in the everyday world. Some would stop all theoretical research, saying that anything but applied research is immoral (Baumrin 1970). Others argue for the value of both basic and applied research while advocating a clear separation between the research and advocate roles (Atkinson 1977). The impatient view that maintains that only applied research is ethical is short-sighted because it fails to recognize that the biggest breakthroughs in science have often come through basic

191

theoretical research. For example, as long as alchemists sought an applied end, to create gold, chemistry did not advance. But once chemists became concerned with understanding chemistry at a theoretical level, there were great leaps in knowledge. The history of other sciences has been similar, suggesting that it would be fool-hardy to abandon the theoretical pursuits that may someday bring about revolutionary solutions to problems.

Scientists who do theoretical work may also try to restrict scientific endeavors to the types of research they conduct and may devalue all applied and action research. This too is a narrow approach to science. The social scientist who is oriented toward social change may apply already-existing knowledge to solving social problems. Also, the social scientist may, by virtue of his training, bring a unique perspective to problem situations and be most aware of the need to empirically evaluate intervention programs. In addition, several eminent scientists have convincingly argued that much applied work also leads to theoretical advances. Thus the two types of research may not be as different as they first appear: applied research often leads to theoretical understanding, and theoretical breakthroughs permit practical applications. Kurt Lewin (1948), a social psychologist, popularized the idea that theoretical advances and the understanding of real human problems should go hand-in-hand.

In applied work, the goals or end product must come from values and considerations that lie outside science. Nevertheless, using the scientific method to find the best way to achieve these goals may also be an excellent way of advancing basic theoretical understanding. Applied social scientists who are working toward certain social goals certainly have a personal right, perhaps even a personal obligation, to do so. However, the partisan scientist, whether working for a liberal cause or for the "establishment," should realize that parti-sanship will place a large strain on his objectivity and should develop attitudes and methodologies to guard against bias. The scientist should consider alternate perspectives and should impartially evalu-ate evidence, even that which disagrees with his approach.

A third possible role for social scientists besides that of theoretical or applied researcher is that of educator—both of students and of the public at large. Not only must we educate the public about the content knowledge of our disciplines (which is still limited) but, perhaps more important, we must teach them a scientific attitude toward problems. Even for problems where social scientists have little more than common sense to offer, the scientist can serve a valuable function by educating those involved about the importance of experimental innovation and empirical evaluation of possible solutions. Scientists should seek to educate others in the questioning and scientific approach to problems exemplified in Webber's (1973, p. 206) description of a scientist:

> But his special character mirrors the special character of science. To a degree far less common in other interest groups, he has learned to *doubt*. He has been trained to question his beliefs, his data, and his findings; to submit his conclusions to critical evaluation by his peers; to tolerate uncertainty and ambiguity; to bear the frustrations of not knowing, and of knowing he does not know; and, by far the most important, to adopt the empirical test for validity.

Whatever role the scientist chooses as being most appropriate personally, he must realize that this is partly a moral choice and commitment. The scientist should realize too that both theoretical and applied work will influence the world and must ethically examine his work in light of the probable outcomes it will produce (see chapter 13).

Conclusion

The reader's own conclusions regarding the relationship between values, objectivity, and science will influence how he does research. If one accepts the value-free approach, he may study problems that are theoretically interesting and have no immediately applied uses. However, one cannot avoid responsibility for knowledge produced. Scientists accepting this approach should be aware that values may creep into the process unnoticed and that value-free research may sometimes serve to support the status quo. One should understand that logic and empiricism can rarely, if ever, guarantee complete objectivity. On the other hand, one may accept a more activist approach to science in which one sets out to do research that will achieve valued personal goals. A scientist adopting this approach to science should realize that it may place strains on his objective exploration of problems. The activist scientist should understand that although logic and empiricism cannot guarantee objectivity, they can greatly increase it. Although a scientist may work in different roles as scientist and as activist, in practice it is difficult to separate oneself into discrete roles. Therefore the individual with an extreme opinion and activist orientation toward a topic will often find it difficult to maintain objectivity. The reader must decide where on the continuum from activist involvement to "arms-length" study he will work. Whatever the decision about this personal choice, the scientist should be aware of his biases, should rely on logical thought and empiricism, and should work to support the "politics of truth."

For a further discussion of values and science see Benne 1965; Black 1977; Bramson 1970; Carovillano and Skehan 1970; Foss 1977; Gouldner 1963; Hudson 1972; Kelman 1968; MacRae 1976; and Sanford 1965.

The Impact of Social Science
on Society: The Scientist's
Responsibility

In a world where anything scientists learn is likely to be put to
immediate and effective use for ends beyond their control and
antithetical to their values, anthropologists must choose their
research undertakings with an eye to their implications. They
must demand the right to have a hand or at least a say in the
use of what they do as a condition for doing it.

Gerald Berreman (1973)

The social scientist today—and particularly the practitioner
and investigator of behavior change—finds himself in a
situation that has many parallels to that of the nuclear
physicist. The knowledge about the control and manipulation
of human behavior that he is producing or applying is beset
with enormous ethical ambiguities, and he must accept
responsibility for its social consequences.

Herbert Kelman (1965)

The present chapter addresses the questions: To what extent are
social scientists responsible for the impact of their science on
society? For what types of knowledge are scientists most directly
responsible? In what concrete ways can they exercise their responsibility? To what extent should behavioral scientists "go public" with
their findings, and when they do so, should they stick to the facts or
also offer informed judgments? Chapter 12 was concerned with the
influence of the investigator's values and beliefs upon science. This
chapter is concerned with the complementary question: What is the
impact of the social sciences on society and who is responsible for
guiding this influence?

The natural and physical sciences have altered civilization. Television, cars, airplanes, the elimination of smallpox, electricity,
atomic energy, and space travel are but a few of the "products" of
science. Within biology the future possibility of genetic transplants
means that we may someday gain the power to control our own

evolution. But other products of the scientific age are napalm, hydrogen bombs, and smog. Modern problems such as the accumulation of nuclear arsenals and growing ecological concerns have forced civilization to realize that the products of science, although undeniably important, are not inevitably beneficial. The tremendous impact of science and the realization that it can be used for either good or evil makes who will control science one of the most important questions of our century.

The impact of the fledgling social sciences has not yet been as revolutionary as that of the older sciences, but the opinion poll, Skinnerian reinforcement, Freudian psychoanalysis, and personality tests are examples of discoveries that have helped shape society. Similarly, IQ tests, encounter groups, the ability to predict elections, and preschool programs have been influential. And there is reason to believe that the products of the social sciences will become more important in the coming years. There will be increasingly effective and controversial "products" that can be used for human control—for example, electrical stimulation of the brain, psychoactive drugs, and behavioral conditioning procedures.

The social sciences will also undoubtedly produce less controversial gifts such as ways of improving parent-child relationships, better methods of education, approaches to lessening prejudice, and ways of decreasing hostility. George Miller (1969, p. 1065) summed up the possibilities when he wrote: "In my opinion, scientific psychology is potentially one of the most revolutionary intellectual enterprises ever conceived by the mind of man. If we were ever to achieve substantial progress toward our stated aim—toward the understanding, prediction, and control of mental and behavioral phenomena—the implications for every aspect of society would make brave men tremble."

The history of the physicial sciences reveals that products of science are not inevitably good. There is no reason to believe this is any different in the behavioral sciences. We must therefore ask, Who will control the products of the behavioral sciences and how can this control be exercised?

Perhaps even more important than the products of social science research are the conceptions of humanity generated by social scientists. George Miller (1969) asserts that when social scientific conceptions about individuals and groups are adopted by the lay public, they have a profound effect on society. The proper way for people to behave and the way problems are defined depends on society's conception of human nature (M. B. Smith 1973), which increasingly comes from the social sciences. And whether the technological products of the social sciences will be implemented will depend on society's definition of what humanity is and what it should be. Thus the ideas about mankind we "give away" will probably ultimately be more important than the technological innovations we produce. Among the important ideas that have already diffused into the cultural mainstream of Western society are a primarily environmental interpretation of most ethnic and cultural differences, a causal approach to deviant behavior, the idea of the importance of childhood in shaping personality, and the ideas of anomie and rootlessness. Just as with the technological products of social science, so too the ideas we propagate may be used for constructive or destructive purposes and so require guidance to insure that they benefit mankind.

The Responsibility of the Social Scientist
The traditional answer to the question of who is responsible for science is that the society in general, and its elected leaders in particular, are to guide the applications of science and technology. The scientist is assigned an active role in the discovery of truth but a passive role in determining societal use of scientific findings. From this vantage point, scientists are seen as technicians, producing knowledge and products for others to use. Many scientists accept this view, maintaining that investigators cannot possibly be held responsible for outcomes they can neither foresee nor control. Besides the idea that the researcher cannot foresee misuses and abuses of scientific findings, the other argument supporting the traditional view is that scientists cannot spend their time in policy debates because they would then not have time for research. In addition, most scientists do not have the skills or the desire to carry out policy-related work such as lobbying. The assumption underlying the traditional view of scientific nonresponsibility is that knowledge is ethically neutral and can be used for good or bad ends. Since political leaders represent the will of the people, it is believed that they insure that scientific discoveries are used for human welfare.

Although scientists still believe in the ultimate right of an informed public to control society, including the applications of scientific knowledge, they now generally take a more active role in the decision process. For example, a much more vocal role was prescribed by the American Association for the Advancement of Science's Committee on Science in the Promotion of Human Welfare (Commoner 1960, p. 71):

> We believe that the scientific community ought to assume, on its own initiative, an *independent* and *active* informative role, whether or not other social agencies see any immediate advantage in hearing what the scientist has to say.... In sum, we conclude that the scientific community should, on its own initiative, assume an obligation to call to public attention those issues of public policy which relate to science, and to provide for the general public the facts and estimates of the effects of alternative policies which the citizen must have if he is to participate intelligently in the solution of these problems. A citizenry thus informed is, we believe, the chief assurance that science will be devoted to the promotion of human welfare.

One reason scientists are now more active in the control and application of their discoveries has been the realization that scientific findings are sometimes misused for political, military, and industrial purposes. Another reason for increased activism is the realization that many of the misuses of science could not have occurred without the help of scientists and that scientists were not mere technicians in these creations but shared a heavy burden of responsibility. The history of Nazi Germany revealed the dangers involved when scientific groups prefer to leave public affairs to others. Scientists also began to realize that they were often in the best position to understand the possible misuses of findings and to predict future uses. The early approach to scientific responsibility restricted the scientist to a conservative and narrow type of involvement in public matters because his major responsibility was believed to be to science. Thus a scientist would be careful to stick to the "facts" (following the customs of scientific discourse) and not make statements that could jeopardize public respect for science. Many scientists now feel a greater responsibility to society and to humanity than to the scientific community (*F.A.S. Public Interest Report* 1976a; Goodfeld 1977) and therefore believe in taking a more active role in policy debates related to science.

198

Predicting the Uses of Knowledge

Although in principle knowledge may be ethically neutral, the uses to which it is likely to be put in the given sociohistorical context will probably not be neutral for human welfare (Kelman 1965). Some scientific findings may seem "neutral" in this respect because it is difficult to tell whether they might be used for constructive or destructive purposes. However, it can be foreseen that certain findings are likely to be used in one way or another. The physicists who developed the nuclear bomb, for example, knew the probable use of their findings and often experienced intense feelings of remorse for the work they had done. As the famous nuclear physicist J. Robert Oppenheimer was made to say in a play about his protest against United States nuclear development: "I begin to wonder whether we were not perhaps traitors to the spirit of science when we handed over the results of our research to the military, without considering the consequences. Now we find ourselves living in a world in which people regard the discoveries of scientists with dread and horror, and go in mortal fear of new discoveries" (*In the Matter of J. Robert Oppenheimer*).

Psychologists and sociologists who have worked on techniques for decreasing racial prejudice (e.g., S. W. Cook 1970) can predict how these findings may be used. As behavioral psychologists developed techniques of behavior modification, they could foresee that they would be used on mental patients and prisoners. Even when research is theoretical in nature, it is often possible to predict how the knowledge will be used. If an anthropologist brings extensive knowledge of a rebellious tribe home to the colonial mother country, he can often predict how this knowledge will be used in the current political climate—whether for greater understanding of the tribe's needs or for carefully calculated maneuvers to suppress insurrection. A sociologist who gathers evidence on the frequency of police brutality for a study on power relationships may predict whether city officials will use this knowledge to suppress brutality. If a psychologist is doing research on the effects of televised violence on real-world behavior, he will assume that these findings can have some influence on television programming.

In many studies, both theoretical and applied, it may happen that the research can benefit one group but be detrimental to other groups or to society as a whole. For example, an industrial

199

psychologist may study factors affecting worker morale in a large company and may foresee, given the current practices and policies of the company, how this knowledge will be used. If the company is exploitative and undemocratic, it is doubtful that workers will be given increased participation in decision-making related to their jobs, based on her findings. Social scientists working for advertising agencies, police departments, and other organizations can often predict the uses that will be made of their research and whether it may help some in society but harm others.

Related to the issue of which group a research project will benefit is a fear that social science research will be destructive because it will be used by the elite or powerful to control less powerful segments of society. For example, industrial firms might use research findings to control their employees, minimize wage demands, and quiet union activists. To insure that social science can be used for the welfare of all of society and not just for elite groups, its findings should be made available to all segments of society, including the powerless and minorities. Research should be conducted not only on the concerns of the powerful (e.g., industry and the military) and the middle-class, but also on the problems and concerns of minorities and of the powerless (cf. Kelman 1978).

It is social scientists' responsibility to make the best possible predictions of how their findings will be used, since they are accountable for the fruits of their labor. If the research is clearly intended to be exploitative it should not be carried out. Often it is possible to insist on a research design that will help protect those who are the focus of the research. The industrial psychologist in the example might insist that the results of her research be open to employees and their union. She might also involve the target group as well as management in the research design and goals and make an effort to see that the changes that result from the research are positive. Usually the choice will not be whether or not to do the research, but how to do the research so the results can benefit both groups.

The more theoretical the research, the harder it usually is to predict how it will eventually be used by society. Kelman (1965) has reviewed the case of basic attitude research and how the findings may help in programs aimed at decreasing prejudice, convincing people to wear seat belts, or adovcating good health practices. But this same research could also be used for propaganda, selling cigarettes, or convincing people to buy things they don't need. Kelman outlined the pros and cons of basic attitude research and showed the moral dilemma often faced in theoretical areas where the eventual applications are unknown.

There is no guarantee that basic theoretical findings will not someday be misused. But there are several ways a scientist can minimize this possibility. One is to disseminate the knowledge to the widest possible audience. In this way there will be some assurance that a powerful group will not use the knowledge simply for its own welfare. Also, if important findings are widely reported, society can insulate itself against misuses.

Another way to prevent abuses is to be alert to destructive applications of theoretical knowledge and work against them. For example, an American anthropologist might report extensive ethnographic data on a native American tribe. If it later appeared that the government might misuse his report to help remove the Indians from their land, it would be incumbent on him to work to prevent this abuse. Scientists must remain alert to misuses, misinterpretations, and overly hasty conclusions based on their findings. Often a scientist must also be involved in translating theoretical work into practical applications, and at this stage he may also exert control and help prevent misuses.

Scientists working in applied areas have an even greater responsibility to work actively to prevent misuse of their findings. For example, behavioral scientists developing new teaching methods, techniques of behavioral control, and effective advertising methods all have a responsibility to evaluate the immediate impact of their research. Similarly, social scientists involved in the areas of program evaluation research and social experimentation have a responsibility to examine the political and other applied implications of their research. Even highly theoretical research can occasionally have an immediate effect on society. As a theory is refined and developed, the scientist must become more alert to who may use the findings and to positive applications of the knowledge. As the theory becomes even more sophisticated, technological offshoots become increasingly likely and at this state the scientist may need to take a more active role in the application of the knowledge.

In summary, at the most theoretical and rudimentary stages of knowledge in an area, it is usually sufficient for scientists to evaluate how this knowledge might someday be used and to present findings in a way designed to forestall misuses. Occasionally the evaluation of uses of knowledge in a developing area will lead the investigator to ask questions in a slightly different way or pursue research that seems more likely to have beneficial potential. As one moves toward the more applied areas and relatively more sophisticated knowledge, scientists must take a more active role in preventing abuses.

Much of the research cited in this book reveals that even theoretical research may have an impact on society. Reiss's study on police brutality, Humphreys's and Kinsey's studies of sexual behavior, Rainwater and Pittman's study of a controversial housing project, and studies of behavior modification are but a few examples of research that may have had an immediate influence on people's lives.

Control and Freedom

One broad concern is that social scientific knowledge may reduce or eliminate human freedom. A disquieting fear has been generated that our growing knowledge of the determinants of human behavior may lead to an end to freedom, a state in which the powerful will use knowledge of behavior to manipulate and control people. Kelman (1965), recognizing that some amount of behavioral control is inevitable at both an individual and a societal level, states that the possibilities of behavioral control are more dangerous than ever before. Society is eager to use behavioral control, and there is growing scientific knowledge of how to control human behavior. Social scientists such as B. F. Skinner (1971) and Hans Eysenck (1969) advocate using social science knowledge to control deviant individuals. Eysenck argues that society can no longer tolerate "social mavericks" and that techniques of psychological conditioning should be used to inculcate "responsible" attitudes in all members of society. See Krapfl and Vargas (1977) and Vargas (1975, 1977) for other behavioristic views.

The total loss of individual freedom would be destructive for a number of reasons. First, there is the ethical dilemma of control per se in a society that highly values freedom. We imprison some people and withdraw their freedom because criminals threaten other values we respect. But in our culture freedom is the ideal state. A second danger is that behavioral control could gradually become widespread, freedom could be slowly lost, and finally the society

202

could be psychologically enslaved, probably without even realizing it. A third danger is that the controllers might misuse their power for their own ends, and history demonstrates that dictators often use every available technique to try to gain complete control. Another possibility is that when social scientists stress the importance of behavioral control rather than concepts like competence and self-regulation, society also comes to think largely in terms of external control of individuals. As G. Miller (1969) said, if scientists' conception of human nature is such that those in power can control people by using scientific information, then government officials will define problems in terms of external control and try to solve them in that way. Even an advocate of behavioral control like B. F. Skinner admits that in the wrong circumstances an effective type of control based upon social scientific knowledge could lead to a terrible "dystopia" rather than the utopia he envisions.

Although control has many negative aspects, in some situations it is highly desirable. For example, learning how to control behaviors such as child abuse, rape, and homicide would be highly desirable. On a less dramatic plane, controlling schizophrenic hallucinations, school truancy, and alcoholism would be very beneficial to society, as would learning to eliminate barriers to racial and sexual equality. There is, then, a conflict between individual freedom and many desirable goals.

How can the conflict between the preservation of freedom and solving problems like crime and prejudice be resolved? Kelman (1965) argues that the behavioral scientist experiences an inevitable tension between these two values and must continually seek a balance. Kelman argues convincingly that although control of others may be dangerous and must be constantly scrutinized, some types of control are much more ethical than others. A number of factors make attempts at behavioral influence or control more or less justifiable from a moral viewpoint: (1) the degree to which the person being influenced can exert a reciprocal influence on the controller; (2) the degree to which the influence benefits the target of the change attempt; and (3) the degree to which the influence attempt will actually enhance the target person's later options. For example, according to Kelman's analysis, when a wife attempts to persuade her husband to exercise more, this probably would be an ethical influence. The appeal is made for his own welfare, the husband may influence her in reciprocal ways (e.g., "You're getting a bit flabby yourself"), and the husband's behavioral options will be greater if he is in good shape.

To the extent that they are successful in understanding and predicting human behavior, the social sciences will also provide opportunities for controlling others. If used judiciously and morally, such control can lead to a better society. For example, if a criminal is taught new prosocial behaviors, many more behavioral options may become available to him. When control attempts are instituted, social scientists must be vigilant to insure that they are constructive. Kelman's three criteria of reciprocity, benefit, and increasing future options provide a way to assess whether control attempts are ethical and to maintain a careful balance between control and individual freedom. Social scientists should also do research on means of resisting destructive control attempts—for instance, ways to neutralize propaganda. Behavioral control will continue to have both positive and negative aspects. Therefore the behavioral scientist must constantly be aware of both its beneficial possibilities in enhancing human welfare and the potential dangers involved.

Methods of Assuming Responsibility

How can scientists exercise responsibility for the findings of their discipline? First of all, they need to speak out publicly. The most famous scientists sometimes enter the public arena to voice the widespread concerns of scientists and to demonstrate that, even with their busy schedules, they will take time to speak out on important issues. Albert Einstein, for example, communicated his beliefs to government officials. Social scientists like Margaret Mead and B. F. Skinner, in addition to their outstanding scientific achievements and active research careers, have found time to speak out clearly on scientific and societal concerns.

One way scientists can collectively influence society's use of their findings is through a division of labor within the scientific community. Some scientists may spend all their time at basic research. But others will have time to write legislators, communicate with the public, and perform other important nonscientific duties. The social sciences include people with diverse temperaments and talents, and some have the interpersonal and communications skills to work competently in the political world. We must actively encourage interested social scientists to represent the opinions of scientists in the social and political arena.

Another way scientists can collectively assume responsibility for the knowledge they generate is through scientific organizations that lobby on their behalf, communicate issues to the public, and provide a forum for the discussion of policy-related issues (Brayfield 1967). In addition to disciplinary societies such as the American Sociological Association, there are organizations that are directed more specifically toward social issues—for example, the Society for Social Responsibility in Science. There are also lobbying associations such as the Association for the Advancement of Psychology. Last, there are interdisciplinary organizations of scientists such as the American Association for the Advancement of Science and the National Academy of Sciences. These numerous associations work to express the beliefs of their members, and active involvement in their decisions is one way scientists who want an active role can influence public policy. In addition to the more permanent associations like those listed above, there often arise loosely federated groups concerned with a specific issue such as the establishment of a federal peace academy, abolishment of capital punishment, or better school programs.

Scientists can take individual responsibility in a variety of ways. They may use the traditional forms of influence in a democracy such as letters to congressmen and to specific agencies. Backed by cogent reasoning and evidence, scientists frequently can influence both the actions of government agencies and the formulation of new legislation. They can write articles about the beneficial and destructive uses of scientific knowledge addressed both to other scientists and to lay persons. Of course many scientists can work for the long-range constructive use of scientific knowledge by their influence on their students. Eminent scientists and experts in specific fields are sometimes called upon for advice by policymakers. Naturally, if one has important knowledge, fears, or strong opinions about an issue, one need not wait to be called on—one can approach policymakers, write position papers, and so forth. In addition, some scientists are involved in policy issues as policy analysts, evaluation researchers, critics of curent practices, and innovators of experimental reforms.

One important, easily overlooked way scientific findings are "misused" is simply that they are not used beneficially when they could be. All the social sciences have accumulated knowledge that is not being put to use. Often this is because policymakers do not know of, or do not understand, the research that would help them formulate better solutions. At other times policymakers ignore social science information because it does not support their political aims. Scientists need to popularize findings and communicate them to policymakers and the public. For example, much of the research on child development and the effects of various child-rearing techniques on personality development has not yet been effectively transmitted to parents. Although there are scattered uses of the findings (as in Parental Effectiveness Training), generally the public has not been adequately educated about them. This failure to use findings in a positive way can be partly alleviated by rewarding social scientists who spend time in communicating information rather than in research and by giving greater recognition to scientists who design applications of theoretical work. Sometimes basic researchers exhibit an unjustified snobbishness and prejudice toward those working in applied areas or directly with the public.

We do not mean to imply that all behavioral scientists should spend a large percentage of their time working in policy areas. But all scientists should at least consider the potential uses of their findings when they plan and present their research. In some cases all scientists have a responsibility to speak out clearly and work actively in relation to specific knowledge areas or issues. But usually most scientists can fulfill their responsibility by spending a small amount of time (and money) in the activities described earlier. Of course we should encourage scientists who do desire to actively represent the behavioral sciences to the public and policymakers. Each scientist must decide for himself the most appropriate and beneficial use of his time.

Politics and Social Science

Closely related to the question of responsibility is the issue of how politically active scientists should be (C. Mills 1959). Since World War II, scientists have taken a much more active role in human affairs. For example, physicists have been vociferous advocates of military arms control, and social scientists have been major defenders of integration and equality in public schools. Thus, questions about scientists' political involvement are important.

To what extent should social scientists, either as individuals or through their professional associations, enter the political arena?

Both sides of the debate agree that a scientist can justifiably speak out on social issues and try to influence the policies of his country: for example, scientists can work for nuclear disarmament. Traditionalists believe the scientist must work for such changes like any other responsible citizen, whereas the "radical" group believes the scientist, as a scientist, has a special knowledge and obligation to work for change—that their associations should speak out on social issues and become politically involved. For example, the American Anthropological Association has issued resolutions pertaining to sex discrimination, the Vietnam War, and fair treatment of Alaskan Eskimos. One ethical principle of the American Anthropological Association (1971, p. 1) reads:

> As people who devote their professional lives to understanding man, anthropologists bear a positive responsibility to speak out publicly both individually and collectively, on what they know and what they believe as a result of their professional expertise gained in the study of human beings. That is, they bear a professional responsibility to contribute to an "adequate definition of reality" upon which public opinion and public policy may be based.

Political issues vary in how closely related they are to the subject matter of particular disciplines. At one end are issues on which social scientists have direct competence or that have direct relevance to the scientific enterprise. An example of the former would be public debates about state mental hospitals, an issue on which clinical psychologists would certainly have special expertise. An issue with direct relevance to the social scientific enterprise would be public hearings on government funding of research. Virtually all researchers would agree that scientists, both as individuals and through their professional associations, should speak out on issues at this end of the continuum. At the other extreme are partisan issues on which social scientists have neither expertise nor a particular vested interest as scientists: for instance, whether there should be agricultural subsidies, whether war resisters should be pardoned, and whether nuclear reactors are likely sources of radioactive pollution.

Although social scientists are usually very concerned about social issues, their opinions often represent their social judgment, not special competence based on scientific findings. Those who oppose

207

social scientific organizations' speaking out on general political issues argue that scientific organizations should limit their recommendations to social issues "about which we are qualified to speak as scientists and professionals" (L. G. Humphreys 1976, p. 6). Those who feel that social scientists should speak out on social issues, regardless of the discipline's scientific knowledge of the area, maintain that scientists and their organizations have political power and therefore have the responsibility to use this power to further humanistic values and policies. Scientists can often make a contribution even in areas where they do not have expert knowledge because they are concerned individuals with a humanistic orientation (Bakan 1976). However, public issues are usually complex, and, even with the best will, scientists are often divided on issues about which there is little scientific consensus. Most may agree that certain problems exist, but in the absence of adequate data, they may disagree greatly about the best solutions. In addition, many issues are political or moral in nature, not scientific. For example, though scientific knowledge does bear on the issue of abortion (e.g. the psychological effects on women who have abortions), the question is largely a moral and legal one.

To make the question of political involvement by social scientists even more complex, social scientific work is actually related to a great many social issues. Between the extreme where the social scientist has obvious expertise and the other extreme where he can only voice his personal opinion are issues about which the social disciplines have some limited and uncertain knowledge—for example, integration, overpopulation, and homosexuality. Weaver (1973, p. 3) suggested that anthropology, and by implication the other social sciences, is political because it acquires knowledge relevant to social problems and social control. He wrote: "Anthropology, by its very existence and by the nature of the data it produces, is a political activity. The American Anthropological Association, by its existence, is a political force. Failure to take a position—or the assumption of a neutral position by failing to face basic ethical problems—is a political action." However, much of our knowledge and theories about many areas are still very uncertain. In addition, issues like integration and abortion are usually value laden, and it is often difficult or impossible to separate the facts a social scientist may testify about from his value orientation.

Traditionalists typically define professional societies as designed to further the progress of science, whereas people most concerned with social action believe they should be groups of individuals with common values and goals. Because nobody has *the truth* about what a professional organization should be, there will continue to be honest-differences of opinion among social scientists about the extent to which their organizations should be politically involved. Partly because of the disagreements about the proper activities of the major scientific societies, other organizations have been founded to work for social change and the application of social scientific knowledge to social problems. Some examples are the Federation of American Scientists, the Society for the Psychological Study of Social Issues, the Society for Social Responsibility in Science, Pugwash, and contributors to the Bulletin of Atomic Scientists.

The activist orientation to social science demands policy statements on social problems by social scientists and their associations, whereas a traditional orientation forbids such activities, except perhaps when a scientist has special expertise on a particular problem. Although there are real differences of opinion about political involvement, most scientists will probably agree with several propositions about political activity. First of all, social scientists collectively have an obligation to produce knowledge, whether theoretical or applied, that may help solve human problems. Second, scientists as individuals must be allowed to be politically active, including using their roles as scientists to work for just causes. Many scientists would also agree that our professional organizations should testify about and become active in social issues about which we have accumulated knowledge. Scientists will differ about whether the social sciences truly have solid knowledge about a particular issue or merely social judgments to offer. However, the leaders who ultimately decide social policy will usually be able to recognize the difference between scientific knowledge and social judgments. Presumably, most politicians understood that when the American Anthropological Association voted to censure the Vietnam War, this was a political and social action and accepted the vote as such. On the other hand, when anthropologists testify about the effect that moving a native American tribe will have on its communal structure, the scientists' testimony is likely to be granted expert status.

Brayfield (1967) and M. B. Smith (1973) have warned that social scientific organizations should be careful not to dilute their voices by becoming involved in every social issue. Social scientific organizations should be most active in areas where they have solid knowledge to contribute, and, as Brayfield clearly warns, social scientists must do their homework and be highly competent if they are to be taken seriously by policymakers. The special role of scientists will always remain that of providing knowledge, including data and principles upon which problem solutions can be built. However, the individual scientist and his professional organization must work to make the implications of this knowledge clear to the public and to policymakers. Whenever scientific organizations work to promote the social judgments of the members, it should be clear that the scientists are speaking only as concerned citizens. As Barnes (1971, p. 342) states, "The members of the Association should not hesitate to state their divergent political views, either collectively or as individuals, but without trying to misappropriate the imprimatur of professional or scientific validity."

Besides facts and judgments, the social scientist is often qualified to contribute additional insights; for example:

1. *Policy analysis.* What alternatives does the public have? What are the likely outcomes of the various alternatives?

2. *Problem definitions.* What are alternative perspectives that would define the problem in new ways?

3. *Solutions.* What new approaches to solving problems can be found?

4. *Program design and evaluation.* How can programs be implemented so that their true effect can be assessed? Are technological innovations possible in this area?

Where political and value judgments are one's major contribution, they should be identified as such so that they do not mislead policymakers or dilute the expert status of behavioral scientists in the areas where we can offer scientific knowledge and expertise.

Several Dilemmas Related to Public Interest

The behavioral scientist working in public policy and political areas related to social science must face a number of dilemmas that require careful judgment (*F.A.S. Public Interest Report* 1976*b*). For example:

1. *When to speak out.* You have apparent knowledge with tentative social and political implications, but years of research are necessary before the knowledge can be confirmed. Should you

present the uncertain findings to the public or to policymakers and suggest their implications?

2. *Providing unsupported opinions.* You have opinions based on general knowledge of behavior and some tangentially related research. Should you speak out on the issue or testify before a legislative body? When you are asked for an unqualified opinion—for "the bottom line"—what should you say?

3. *The problem of allies.* Legislators and other policymakers who support your opinion present it in an inaccurate and exaggerated way. What do you do?

4. *Certainty of expression.* Should your public statements be the highly cautious ones that characterize scientific discourse, or should they convey the greater certainty that you personally feel and that is more likely to convince policymakers?

5. *Going public.* When should you "go public" and appeal to the citizens to support your opinion? When should findings that have not been published in scientific journals and therefore have not been critically examined by colleagues be presented in the popular media? When are you justified in presenting a dramatic and compelling case for your side to the public (without the usual scientific limitations spelled out)?

Of course these questions cannot be easily answered in the abstract; even in specific cases they are difficult to answer. They are dilemmas because there are positive and negative aspects to whatever choices one makes.

Summary

There is every reason to believe that the social sciences will have a profound effect on society. The natural and physical sciences have been revolutionary in their influence, completely changing civilization. Although the social sciences possess much less sophisticated knowledge, they have already begun to affect society and this will undoubtedly increase in the coming decades. One powerful influence of the social sciences is that they may greatly shape humanity's conception of itself by the ideas they "give away" to the public.

How is the impact of science to be regulated? The traditional view was that the leaders of society should make judgments about the application of scientific findings. But there is now an increased emphasis on the responsibility of scientists in these decisions.

Scientists need to act to make knowledge available to all segments of the public and to point out possible abuses of science. They and their associations should work for the constructive use of science and to prevent abuses such as have occurred in the past. The uses of applied research are usually obvious, whereas the uses to which theoretical knowledge may be put are often difficult to predict. Nevertheless, social scientists as a group need to promote uses of scientific knowledge, both theoretical and applied, that will benefit human welfare. Scientists must be alert for two dangers in the application of their findings: Will they be used to limit people's freedom in an unjustified and dangerous way? Will only the powerful or elite profit from the application of the research? Both possibilities may be lessened if scientists are vigilant and if they do research on the concerns of all segments of society.

At both the individual and the collective level scientists can offer many things besides the "facts"—for example, expert judgments, an analysis of alternative courses of action, new problem definitions and solutions, and advice on program evaluation. In addition, the scientist may offer social and value judgments. Since it is usually impossible to completely separate the roles of scientist and concerned citizen, it is also impossible to specify just what a scientist may and may not convey in policy debates. The important thing is to make explicit the basis of whatever ideas the scientist offers, so that statements of opinion, researched fact, judgment, and values can be differentiated by policymakers and the public.

For a further discussion of responsibility for the impact of science on society, see Caplan and Nelson 1973; Kelman 1965; Mead 1962; G. A. Miller 1969; M. B. Smith 1969, 1974; and Vallance 1972.

V Ethical Suggestions

14 Guidelines

The following brief suggestions are meant not to be a summary of the material covered in the book, but to convey some of the general guidelines that emerge.

Sensitivity and Responsibility

Ethical decisions are made by concerned and knowledgeable people who realize the value implications of their choices. The ethical researcher is concerned about the well-being of research participants and about the future uses of the knowledge, and he accepts personal responsibility for decisions bearing on them. The basic ethical imperatives are that the scientist be concerned about the welfare of subjects, be knowledgeable about issues of ethics and values, take these into account when making research decisions, and accept responsibility for his decisions and actions.

Precautions to Safeguard Participants

Research participants must be protected. In research exposing subjects to considerable risk, their safety must be insured by stringent safeguards, including carefully selecting subjects and checking afterward for harmful effects. The scientist has a positive obligation to correct any harm that does befall a participant.

Informed Consent

If the participants will be deprived of rights or exposed to serious risks, they should be informed of this and allowed to withdraw from the research. Subjects should usually be informed beforehand about aspects of the study that would affect their decision on whether to participate. They have the right to withdraw at any time, and strong pressure should not be used to gain cooperation. Even when no dangers are inherent in the research, subjects value informed consent, and it should be obtained whenever feasible.

215

The Less Powerful

Special care should be taken to protect the rights and interests of the less powerful participants in research such as children, the poor, minorities, prisoners, and patients. Scientists should conduct more research on the concerns and needs of the less powerful and consider their viewpoints when formulating studies.

Privacy

Very private information about participants may be collected only with their consent. All research information on individuals should be strictly confidential and published only in summary form unless participants agree that they may be named in the report.

Deception

Research deceptions should never be practiced until an ethical analysis of the situation has been made. Are there other ways to obtain the knowledge? What will be the negative effects of the deception? Can safeguards such as forewarnings and debriefing be used? Deceptions vary from mild to blatant, and, though many mild deceptions may be justifiable, large deceptions often are not. In addition to the ethical questions, deception research often suffers from methodological problems. Also, use of gross deceptions might seriously damage society's respect for the social sciences.

Review by Others

If the investigator is unsure about the ethics of his research, he should seek the opinions of others. If subjects are to be exposed to risks or if the research raises serious value questions, it is wise to solicit the opinions of several reviewers. Disinterested persons may have a sounder ethical perspective than the scientist who is deeply involved with the research. It is often important to gain input from participants as well as from professional colleagues.

Experiments in Change

When research is directed at changing individuals or a group, those who are the target of change should be consulted and their wishes and needs respected. Usually the target group can be involved in setting the goals of the change attempt. When various treatment groups are used in formal experiments, the scientist should carefully consider whether the various experimental manipulations are ethi-

216

cal. A group should not be placed at a serious disadvantage unless this possibility has been accepted by subjects or unless resources are insufficient to offer the most desirable treatment to all persons. Often the treatment that is found most effective can be offered to all participants after the study.

Objectivity and Competence

Complete scientific objectivity is an ideal that cannot be realized in practice, but the scientist should strive to be as objective as possible in conducting research. Biases should never be deliberately introduced into the design or reporting of studies. Since poor research based on faulty methodology and design does not advance knowledge and wastes valuable resources, all social scientists have a responsibility to do the best research they are capable of. Results should be reported accurately and honestly, without omissions that would seriously affect their interpretation. Although values may influence the topic of research, the methodology should be designed to advance truth and not simply support a predetermined position.

Uses of Research Knowledge

Social scientists are responsible for how their discoveries are used. When a study is supported by a funding agency, the scientist must determine whether the research will be used for beneficial purposes. He should examine the possible applications of social scientific findings and endeavor to make these uses constructive. Before conducting a study the researcher must consider how the information will affect the people being studied. At a more general level, scientists have an obligation to speak out individually and collectively when they possess expert knowledge that bears on important societal issues.

Appendixes

Appendix A
CODES OF ETHICS

Included here are selected portions of the codes of ethics of the American Anthropological Association, American Psychological Association, and American Sociological Association. The full codes of ethics of these and other professional associations may be acquired by writing directly to the associations.

American Anthropological Association
PRINCIPLES OF PROFESSIONAL RESPONSIBILITY
Preamble
Anthropologists work in many parts of the world in close personal association with the peoples and situations they study. Their professional situation is, therefore, uniquely varied and complex. They are involved with their discipline, their colleagues, their students, their sponsors, their subjects, their own and host governments, the particular individuals and groups with whom they do their field work, other populations and interest groups in the nations within which they work, and the study of processes and issues affecting general human welfare. In a field of such complex involvements, misunderstandings, conflicts, and the necessity to make choices among conflicting values are bound to arise and to generate ethical dilemmas. It is a prime responsibility of anthropologists to anticipate these and to plan to resolve them in such a way as to do damage neither to those whom they study nor, in so far as possible, to their scholarly community. Where these conditions cannot be met, the anthropologist would be well-advised not to pursue the particular piece of research.

The following principles are deemed fundamental to the anthropologist's responsible, ethical pursuit of his profession.

1. *Relations with those studied:*
In research, an anthropologist's paramount responsibility is to those he studies. When there is a conflict of interest, these in-

221

dividuals must come first. The anthropologist must do everything within his power to protect their physical, social, and psychological welfare and to honor their dignity and privacy.

a. Where research involves the acquisition of material and information transferred on the assumption of trust between persons, it is axiomatic that the rights, interests, and sensitivities of those studied must be safeguarded.

b. The aims of the investigation should be communicated as well as possible to the informant.

c. Informants have a right to remain anonymous. This right should be respected both where it has been promised explicitly and where no clear understanding to the contrary has been reached. These strictures apply to the collection of data by means of cameras, tape recorders, and other data-gathering devices, as well as to data collected in face-to-face interviews, or in participant observation. Those being studied should understand the capacities of such devices; they should be free to reject them if they wish; and if they accept them, the results obtained should be consonant with the informant's right to welfare, dignity, and privacy.

d. There should be no exploitation of individual informants for personal gain. Fair return should be given them for all services.

e. There is an obligation to reflect on the foreseeable repercussions of research and publication on the general population being studied.

f. The anticipated consequences of research should be communicated as fully as possible to the individuals and groups likely to be affected.

g. In accordance with the Association's general position on clandestine and secret research, no reports should be provided to sponsors that are not also available to the general public and, where practicable, to the population studied.

h. Every effort should be exerted to cooperate with members of the host society in the planning and execution of research projects.

i. All of the above points should be acted upon in full recognition of the social and cultural pluralism of host societies and the consequent plurality of values, interests, and demands in those societies. This diversity complicates choice-making in research, but ignoring it leads to irresponsible decisions.

2. *Responsibility to the public:*

The anthropologist is also responsible to the public—all presumed consumers of his professional efforts. To them he owes a commitment to candor and to truth in the dissemination of his research results and in the statement of his opinions as a student of man.

a. He should not communicate his feelings secretly to some and withhold them from others.

b. He should not knowingly falsify or color his findings.

c. In providing professional opinions, he is responsible not only for their content but also for integrity in explaining both these opinions and their bases.

d. As people who devote their professional lives to understanding man, anthropologists bear a positive responsibility to speak out publicly, both individually and collectively, on what they know and they believe as a result of their professional expertise gained in the study of human beings. That is, they bear a professional responsibility to contribute to an "adequate definition of reality" upon which public opinion and public policy may be based.

e. In public discourse, the anthropologist should be honest about his qualifications and cognizant of the limitations of anthropological expertise.

3. *Responsibility to the discipline:*

An anthropologist bears responsibility for the good reputation of his discipline and its practitioners.

a. He should undertake no secret research, or any research whose results cannot be freely derived and publicly reported.

b. He should avoid even the appearance of engaging in clandestine research, by fully and freely disclosing the aims and sponsorship of all his research.

c. He should attempt to maintain a level of integrity and rapport in the field such that by his behavior and example he will not jeopardize future research there. The responsibility is not to analyze and report so as to offend no one, but to conduct research in a way consistent with a commitment to honesty, open inquiry, clear communication of sponsorship and research aims, and concern for the welfare and privacy of informants.

4. *Responsibility to students:*

In relations with students an anthropologist should be candid, fair, nonexploitative and committed to their welfare and academic progress....

b. He should alert students to the ethical problems of research and discourage them from participating in projects employing questionable ethical standards. This should include providing them with information and discussions to protect them from unethical pressures and enticements emanating from possible sponsors, as well as helping them to find acceptable alternatives....

g. He should acknowledge in print the student assistance he uses in his own publications, give appropriate credit (including coauthorship) when student research is used in publication, encourage and assist in publication of worthy student papers, and compensate students justly for the use of their time, energy, and intelligence in research and teaching....

5. *Responsibility to sponsors:*

In his relations with sponsors of research, an anthropologist should be honest about his qualifications, capabilities, and aims. He thus faces the obligation, prior to entering any commitment for research, to reflect sincerely upon the purposes of his sponsors in terms of their past behavior. He should be especially careful not to promise or imply acceptance of conditions contrary to his professional ethics or competing commitments. This requires that he require of the sponsor full disclosure of the sources of funds, personnel, aims of the institution and the research project, disposition of the research results. He must retain the right to make all ethical decisions in his research. He should enter no secret agreement with the sponsor regarding the research, results, or reports.

6. *Responsibilities to one's own government and to host governments:*

In his relation with his own government and with host governments, the research anthropologist should be honest and candid. He should demand assurance that he will not be required to compromise his professional responsibilities and ethics as a condition of his permission to pursue the research. Specifically, no secret research, no secret reports or debriefing of any kind should be agreed to or given. If these matters are clearly understood in advance, serious complications and misunderstandings can generally be avoided.

Epilogue:

In the final analysis, anthropological research is a human undertaking dependent upon choices for which the individual bears ethical as well as scientific responsibility. That responsibility is a human, not superhuman responsibility. To err is human, to forgive humane. This statement of principles of professional responsibility is not designed to punish, but to provide guidelines which can minimize the occasions upon which there is a need to forgive. When an anthropologist, by his actions, jeopardizes peoples studied, professional colleagues, students or others, or if he otherwise betrays his professional commitments, his colleagues may legitimately inquire into the propriety of those actions, and take such measures as lie within the legitimate powers of their Association as the membership of the Association deems appropriate.

Adopted May, 1971

American Psychological Association
ETHICAL PRINCIPLES IN THE CONDUCT OF RESEARCH WITH HUMAN PARTICIPANTS
The decision to undertake research should rest upon a considered judgment by the individual psychologist about how best to contribute to psychological science and to human welfare. The responsible psychologist weighs alternative directions in which personal energies and resources might be invested. Having made the decision to conduct research, psychologists must carry out their investigations with respect for the people who participate and with concern for their dignity and welfare. The Principles that follow make explicit the investigator's ethical responsibilities toward participants over the course of research, from the initial decision to pursue a study to the steps necessary to protect the confidentiality of research data. These Principles should be interpreted in terms of the context provided in the complete document offered as a supplement to these Principles.

1. In planning a study the investigator has the personal responsibility to make a careful evaluation of its ethical acceptability, taking into account these Principles for research with human beings. To the extent that this appraisal, weighing scientific and humane values, suggests a deviation from any Principle, the investigator incurs an increasingly serious obligation to seek ethical advice and to observe more stringent safeguards to protect the rights of the human research participant.

2. Responsibility for the establishment and maintenance of acceptable ethical practice in research always remains with the individual investigator. The investigator is also responsible for the ethical treatment of research participants by collaborators, assistants, students, and employees, all of whom, however, incur parallel obligations.

225

3. Ethical practice requires the investigator to inform the participant of all features of the research that reasonably might be expected to influence willingness to participate and to explain all other aspects of the research about which the participant inquires. Failure to make full disclosure gives added emphasis to the investigator's responsibility to protect the welfare and dignity of the research participant.

4. Openness and honesty are essential characteristics of the relationship between investigator and research participant. When the methodological requirements of a study necessitate concealment or deception, the investigator is required to ensure the participant's understanding of the reasons for this action and to restore the quality of the relationship with the investigator.

5. Ethical research practice requires the investigator to respect the individual's freedom to decline to participate in research or to discontinue participation at any time. The obligation to protect this freedom requires special vigilance when the investigator is in a position of power over the participant. The decision to limit this freedom increases the investigator's responsibility to protect the participant's dignity and welfare.

6. Ethically acceptable research begins with the establishment of a clear and fair agreement between the investigator and the research participant that clarifies the responsibilities of each. The investigator has the obligation to honor all promises and commitments included in that agreement.

7. The ethical investigator protects participants from physical and mental discomfort, harm and danger. If the risk of such consequences exists, the investigator is required to inform the participant of that fact, to secure consent before proceeding, and to take all possible measures to minimize distress. A research procedure may not be used if it is likely to cause serious and lasting harm to participants.

8. After the data are collected, ethical practice requires the investigator to provide the participant with a full clarification of the nature of the study and to remove any misconceptions that may have arisen. Where scientific or humane values justify delaying or withholding information, the investigator acquires a special responsibility to assure that there are no damaging consequences for the participant.

9. Where research procedures may result in undesirable consequences for the participant, the investigator has the responsibility to
detect and remove or correct these consequences, including, where
relevant, long-term aftereffects.

10. Information obtained about the research participants during
the course of an investigation is confidential. When the possibility
exists that others may obtain access to such information, ethical
research practice requires that this possibility, together with the
plans for protecting confidentiality, be explained to the participants
as a part of the procedure for obtaining informed consent.

Adopted December, 1972

American Sociological Association

Preamble

Sociological inquiry is often disturbing to many persons and groups.
Its results may challenge long-established beliefs and lead to change
in old taboos. In consequence such findings may create demands
for the supression or control of this inquiry or for a dilution of the
findings. Similarly, the results of sociological investigation may be of
significant use to individuals in power—whether in government, in
the private sphere, or in the universities—because such findings,
suitably manipulated, may facilitate the misuse of power. Knowledge is a form of power, and in a society increasingly dependent on
knowledge, the control of information creates the potential for
political manipulation.

For these reasons, we affirm the autonomy of sociological inquiry.
The sociologist must be responsive, first and foremost, to the truth
of his investigation. Sociology must not be an instrument of any
person or group who seeks to suppress or misuse knowledge. The
fate of sociology as a science is dependent upon the fate of free
inquiry in an open society.

At the same time this search for social truths must itself operate
within constraints. Its limits arise when inquiry infringes on the
rights of individuals to be treated as persons, to be considered—in
the renewable phrase of Kant—as ends and not means. Just as
sociologists must not distort or manipulate truth to serve untruthful
ends, so too they must not manipulate persons to serve their quest
for truth. The study of society, being the study of human beings,
imposes the responsibility of respecting the integrity, promoting the

dignity, and maintaining the autonomy of these persons.

To fulfill these responsibilities, we, the members of the American Sociological Association, affirm the following Code of Ethics:

Code of Ethics

1. *Objectivity in Research*

In his research the sociologist must maintain scientific objectivity.

2. *Integrity in Research*

The sociologist should recognize his own limitations and, when appropriate, seek more expert assistance or decline to undertake research beyond his competence. He must not misrepresent his own abilities, or the competence of his staff to conduct a particular research project.

3. *Respect of the Research Subject's Rights to Privacy and Dignity*

Every person is entitled to the right of privacy and dignity of treatment. The sociologist must respect these rights.

4. *Protection of Subjects from Personal Harm*

All research should avoid causing personal harm to subjects used in research.

5. *Preservation of Confidentiality of Research Data*

Confidential information provided by a research subject must be treated as such by the sociologist. Even though research information is not a privileged communication under the law, the sociologist must, as far as possible, protect subjects and informants. Any promises made to such persons must be honored. However, provided that he respects the assurances he has given his subjects, the sociologist has no obligation to withhold information of misconduct of individuals or organizations.

If an informant or other subject should wish, however, he can formally release the researcher of a promise of confidentiality. The provisions of this section apply to all members of research organizations (i.e., interviewers, coders, clerical staff, etc.), and it is the responsibility of the chief investigators to see that they are instructed in the necessity and importance of maintaining the confidentiality of the data. The obligation of the sociologist includes the use and storage of original data to which a subject's name is attached. When requested, the identity of an organization or subject must be adequately disguised in publication.

6. *Presentation of Research Findings*

The sociologist must present his findings honestly and without distortion. There should be no omission of data from a research report which might significantly modify the interpretation of findings.

7. *Misuse of Research Role*

The sociologist must not use his role as a cover to obtain information for other than professional purposes.

8. *Acknowledgment of Research Collaboration and Assistance*

The sociologist must acknowledge the professional contributions or assistance of all persons who collaborated in the research.

9. *Disclosure of the Sources of Financial Support*

The sociologist must report fully all sources of financial support in his research publications and any special relations to the sponsor that might affect the interpretation of the findings.

10. *Distortion of Findings by Sponsor*

The sociologist is obliged to clarify publicly any distortion by a sponsor or client of the findings of a research project in which he has participated.

11. *Disassociation from Unethical Research Arrangements*

The sociologist must not accept such grants, contracts, or research assignments as appear likely to require violation of the principles above, and must publicly terminate the work or formally disassociate himself from the research if he discovers such a violation and is unable to achieve its correction.

12. *Interpretation of Ethical Principles*

When the meaning and application of these principles are unclear, the sociologist should seek the judgment of the relevant agency or committee designated by the American Sociological Association. Such consultation, however, does not free the sociologist from his individual responsibility for decisions or from his accountability to the profession.

13. *Applicability of Principles*

In the conduct of research the principles enunciated above should apply to research in any area either within or outside the United States of America.

14. *Interpretation and Enforcement of Ethical Principles*

The Standing Committee on Professional Ethics, appointed by the Council of the Association, shall have primary responsibility for the interpretation and enforcement of the Ethical Code....

Effective September 1, 1971

229

Animal Experimentation

See: *Guide for the care and use of laboratory animals* 1974; Guidelines for the use of animals in school science behavior projects 1972; *F.A.S. Public Interest Report* 1977. The reader should visit a local institutional research office to obtain information regarding current laws governing the care and use of laboratory animals. Or in the United States, write to the Institute of Laboratory Animal Resources, National Institutes of Health.

Laws Governing Research with Humans

We have not discussed specific laws governing research because readers will be from many countries and from different legal jurisdictions. In addition, since laws within the United States are currently developing very rapidly, it is impossible to refer to laws that would certainly apply. Instead, we recommend checking with a local institutional research office or a major research branch of your government. Within the United States a helpful source of information is the National Commission for the Protection of Human Subjects of Biomedical and Behavioral Research, U.S. Department of Health, Education, and Welfare.

For several references related to behavioral research and the law, see Annas, Glantz, and Katz 1977; Berry, Castaneda, and Morton 1970; Freund 1967; Nejelski 1976; Silverman 1975; Weinberger 1974, 1975.

Bibliographies

General bibliographies related to ethics and values may be found in Bureau of Social Science Research 1975; and West 1973.

Books

Several books related to the ethical and value issues of research are American Psychological Association 1973; Annas, Glantz, and

Further References

Katz 1977; Arellano-Galdames 1973; Barber et al. 1973; Benne 1965; Carovillano and Skehan 1970; Denzin 1973; Gray 1975; Hook, Kurtz, and Tŏdŏrŏvich 1977; Horowitz 1967; Katz 1972; Kelman 1968; Kennedy 1975; MacRae 1976; A. R. Miller 1971; National Academy of Sciences 1975; Nejelski 1976; Rivlin and Timpane 1975; Rynkiewich and Spradley 1976; Sjoberg 1971; Smith 1969; Weaver 1973; Westin 1967.

References

Abrahams, D. 1967. The effect of concern on debriefing following a deception experiment. Master's thesis, University of Minnesota. Cited in Walster, E.; Berscheid, E.; Abrahams, D.; and Aronson, V. *Journal of Personality and Social Psychology* 6:371–80.

Adams, R. 1971. Responsibilities of the foreign scholar to the local scholarly community. *Current Anthropology* 12 (June): 335–39.

Adorno, T. W.; Frenkel-Brunswik, E.; Levinson, D. J.; and Sanford, R. N. 1950. *The authoritarian personality.* New York: Harper.

Allen, V. 1966. Effect of knowledge of deception on conformity. *Journal of Social Psychology* 69:101–6.

American Anthropological Association. 1971. *Principles of professional responsibility.* Washington, D.C.: American Anthropological Association.

American Psychological Association. 1973. *Ethical principles in the conduct of research with human subjects.* Washington, D.C.: Ad hoc Committee on Ethical Standards in Psychological Research, American Psychological Association.

American Sociological Association. 1971. *Code of ethics.* Washington, D.C.: American Sociological Association.

————. 1977. ASA testimony before the Commission for the Protection of Human Subjects. *ASA Footnotes* 5:5, 9.

Amrine, M., ed. 1965. Testing and public policy. *American Psychologist* 20:857–993.

Amrine, M., and Sanford, F. H. 1956. In the matter of juries, democracy, science, truth, senators and bugs. *American Psychologist* 11:54–60.

Annas, G. J. 1976. Confidentiality and the duty to warn. *Hastings Center Report* 6:6–8.

Annas, G. J.; Glantz, L. H.; and Katz, B. F. 1977. *Informed consent to human experimentation: The subject's dilemma.* Cambridge, Mass.: Ballinger.

Arellano-Galdames, F. 1973. Some ethical problems in research on human subjects. *Dissertation Abstracts* 33, 12-A (June): 6579.

Ares, C. E.; Rankin, I.; and Sturz, H. 1973. The Manhattan bail project: An interim report on the use of pretrial parole. *New York University Law Review* 38:67–95.

233

Argyris, C. 1968. Some unintended consequences of rigorous research. *Psychological Bulletin* 70:185-97.

Aronson, E. 1966. Avoidance of inter-subject communication. *Psychological Reports* 19:238.

Aronson, E., and Carlsmith, J. M. 1968. Experimentation in social psychology. In *Handbook of social psychology,* ed. G. Lindzey and E. Aronson, vol. 2. Reading, Mass.: Addison-Wesley.

Asher, J. 1974. Can parapsychology weather the Levy affair? *APA Monitor* 5:4.

Association of Black Psychologists. 1975. On the report of the Ad Hoc Committee on Educational Uses of Tests with Disadvantaged Students. *American Psychologist* 30:88-95.

Astin, A. W., and Boruch, R. F. 1970. A "link" system for assuring confidentiality of research data in longitudinal studies. *American Educational Research Journal* 7:615-24.

Atkinson, R. C. 1977. Reflections on psychology's past and concerns about its future. *American Psychologist* 32:205-10.

Ax, A. F. 1953. The physiological differentiation between fear and anger in humans. *Psychosomatic Medicine* 15:433-42.

Azrin, N. H.; Holz, W.; Ulrich, R.; and Goldiamond, I. 1961. The control of the content of conversation through reinforcement. *Journal of the Experimental Analysis of Behavior* 4:25-30.

Bakan, D. 1976. Psychology and public policy. Unpublished paper, York University, Ontario, Canada.

Baldwin, A. L. 1949. The effect of home environment on nursery school behavior. *Child Development* 20:49-62.

Barber, B. 1962. *Science and the social order.* New York: Collier Books.

————. 1975. Some perspectives on the role of assessment of risk/benefit criteria in the determination of the appropriateness of research involving human subjects. Paper prepared for the National Commission for the Protection of Human Subjects of Biomedical and Behavioral Research, U.S. Department of Health, Education, and Welfare, Bethesda, Md.

Barber, B.; Lally, J. J.; Makarushka, J. L.; and Sullivan, D. 1973. *Research on human subjects: Problems of social control in medical experimentation.* New York: Russell Sage Foundation.

Barber, T. X. 1973. Pitfalls in research: Nine investigator and experimenter artifacts. In *Second handbook of research on teaching,* ed. R. M. W. Travers. Chicago: Rand McNally.

————. 1976. *Pitfalls in human research.* New York: Pergamon.

Barnes, J. A. 1963. Some ethical problems in modern fieldwork. *British Journal of Sociology* 14:118-34.

————. 1971. Comments. *Current Anthropology* 12:342.

Baron, R. A., and Eggleston, R. J. 1972. Performance on the "aggression machine": Motivation to help or harm? *Psychonomic Science* 26:321-22.

References

Baumrin, B. H. 1970. The immorality of irrelevance: The social role of science. In *Psychology and the problems of society,* ed. F. F. Korten, S. W. Cook, and J. I. Lacey. Washington, D.C.: American Psychological Association.

Baumrind, D. 1964. Some thoughts on ethics of research: After reading Milgram's behavioral study of obedience. *American Psychologist* 19:421-23.

————. 1971. Principles of ethical conduct in the treatment of subjects: Reaction to the draft report of the Committee on Ethical Standards in Psychological Research. *American Psychologist* 26:887-96.

————. 1976. Nature and definition of informed consent in research involving deception. Paper prepared for the National Commission for the Protection of Human Subjects of Biomedical and Behavioral Research, U.S. Department of Health, Education, and Welfare, Bethesda, Md.

Beals, R. L. 1967. Cross cultural research and government policy. *Bulletin of the Atomic Scientists* 23:18-24.

————. 1969. *Politics of social research.* Chicago: Aldine.

Becker, H. S. 1967. Whose side are we on? *Journal of Social Problems* 14:239-47.

Becker, H. S., and Geer, B. 1957. Participant observation and interviewing: A comparison. *Human Organization* 16:28-32.

Beecher, H. K. 1966a. Documenting the abuses. *Saturday Review* 49:45-46.

————. 1966b. Ethics and clinical research. *New England Journal of Medicine* 274:1354-60.

————. 1970. *Research and the individual: Human Studies.* Boston: Little, Brown.

Benne, K. D., 1965. The social responsibilities of the behavioral scientist. *Journals of Social Issues*, vol. 21, no. 2 (April), entire issue.

Bennett, C. C. 1967. What price privacy? *American Psychologist* 22:371-76.

Berkowitz, L. 1976. Some complexities and uncertainties regarding the ethicality of deception in research with human subjects. Paper prepared for the National Commission for the Protection of Human Subjects of Biomedical and Behavioral Research, U.S. Department of Health, Education, and Welfare, Bethesda, Md.

Berkun, M.; Bialek, H. M.; Kern, P. R.; and Yagi, K. 1962. Experimental studies of psychological stress in man. *Psychological Monographs: General and Applied* 76 (15): 1-39.

Berreman, G. D. 1969. Academic colonialism: Not so innocent abroad. *Nation* 209:505-8.

————. 1973. The social responsibility of the anthropologist. In *To see ourselves: Anthropology and modern social issues,* ed. T. Weaver. Glenview, Ill.: Scott, Foresman.

Berry, R. G.; Castaneda, A.; and Morton, D. 1970. Psychology and the law:

A symposium. *Canadian Psychologist* 11:2–29.

Berscheid, E.; Baron, R. S.; Dermer, M.; and Libman, M. 1973. Anticipating informed consent. *American Psychologist* 28:913–25.

Bickman, L., and Henchy, T. 1972. *Beyond the laboratory: Field research in social psychology.* New York: McGraw-Hill.

Black, M. 1977. The objectivity of science. *Bulletin of the Atomic Scientists.* 33:55–60.

Boas, F. 1919. Correspondence: Scientists as spies. *Nation* 109:797.

Bonacich, P. 1970. Deceiving subjects: The pollution of our environment. *American Sociologist* 5:45.

Borek, E. 1976. The loneliness of the original investigator. *Nature* 264:100.

Boruch, R. F. 1971a. Assuring confidentiality of responses in social research: A note on strategies. *American Sociologist* 6:308–11.

———. 1971b. Maintaining confidentiality of data in education research: A systemic analysis. *American Psychologist* 26:413–30.

———. 1972. Relations among statistical methods for assuring confidentiality of social research data. *Social Science Research* 1:403–14.

———. 1976. Strategies for eliciting and merging confidential social research data. In *Social research in conflict with law and ethics,* ed. P. Nejelski. Cambridge, Mass.: Ballinger.

Bramson, L. 1970. Reflections on the social and political responsibility of sociologists. In *The sociology of sociology,* ed. P. Halmos. Keele: University of Keele.

Brayfield, A. H. 1967. Psychology and public affairs. *American Psychologist* 22:182–86.

Brock, T. C., and Becker, L. A. 1966. Debriefing and susceptibility to subsequent experimental manipulations. *Journal of Experimental Social Psychology* 2:314–23.

Bronfenbrenner, U. 1959. Freedom and responsibility in research: Comments. *Human Organization* 18:49–52.

———. 1973. *Two worlds of childhood: U.S. and U.S.S.R.* New York: Pocket Books.

Brower, D. 1948. The role of incentive in psychological research. *Journal of General Psychology* 39:145–47.

Brown, R. 1965. *Social psychology.* New York: Free Press.

Bryan, J. H., and Test, M. A. 1967. Models and helping: Naturalistic studies in aiding behavior. *Journal of Personality and Social Psychology* 6:400–407.

Bryant, E. C., and Hansen, M. H. 1976. Invasion of privacy and surveys: A growing dilemma. In *Perspectives on attitude assessment: Surveys and their alternatives,* ed. H. W. Sinaiko and L. A. Broedling. Champaign, Ill.: Pendleton Publications.

Burchard, W. W. 1957. A study of attitudes toward the use of concealed devices in social science research. *Social Forces* 36:111–16.

References

Bureau of Social Science Research. 1975. *The ethics of social research—a selected and annotated bibliography.* Washington, D.C.: Bureau of Social Science Research.

Buros, O. K. 1972. *Mental measurements yearbook.* Highland Park, N.J.: Gryphon Press.

———. 1974. *Tests in print.* Highland Park, N.J.: Gryphon Press.

Campbell, D.; Sanderson, R. E.; and Laverty, S. G. 1964. Characteristics of a conditioned response in human subjects during extinction trials following a single traumatic conditioning trial. *Journal of Abnormal and Social Psychology* 68:627-39.

Campbell, D. T. 1969a. Prospective: Artifact and control. In *Artifact in behavioral research,* ed. R. Rosenthal and R. L. Rosnow. New York: Academic Press.

———. 1969b. Reforms as experiments. *American Psychologist* 24:409-29.

———. 1976. Protection of the rights and interests of human subjects in program evaluation, social indicators, social experimentation, and statistical analysis based upon administrative records. Paper prepared for the National Commission for the Protection of Human Subjects of Biomedical and Behavioral Research, U.S. Department of Health, Education, and Welfare, Bethesda, Md.

Campbell, D. T., and Stanley, J. 1967. *Experimental and quasi-experimental designs for research.* Chicago: Rand McNally.

Caplan, N., and Nelson, S. D. 1973. On being useful: The nature and consequences of psychological research. *American Psychologist* 28:199-211.

Carlson, R. 1971. Where is the person in personality research? *Psychological Bulletin* 75:203-19.

Carovillano, R. L., and Skehan, J. W. 1970. *Science and the future of man.* Cambridge, Mass.: MIT Press.

Carroll, J. D. 1973. Confidentiality of social science research sources and data: The Popkin case. *PS Newsletter* 6:268-80.

Chambliss, W. 1975. On the paucity of original research on organized crime. *American Sociologist* 10:36-39.

Christie, R., and Geis, F. L. 1970. *Studies in Machiavellianism.* New York: Academic Press.

Clark, R. D., and Word, L. E. 1974. Where is the apathetic bystander? Situational characteristics of the emergency. *Journal of Personality and Social Psychology* 29:279-87.

Commoner, B., et al. 1960. AAAS Committee on Science in the Promotion of Human Welfare states the issues and calls for action. *Science* 132: 68-73.

Conrad, H. S. 1967. Clearance of questionnaires with respect to invasion of privacy, public sensitivities, ethical standards, etc. *American Psychologist* 22:356-59.

Cook, S. W. 1970. Motives in a conceptual analysis of attitude related behavior. In *Nebraska Symposium on Motivation,* 1969, ed. W. J. Arnold and D. Levine. Lincoln: University of Nebraska Press.

————. 1975. A comment on the ethical issues involved in West, Gunn, and Chernicky's "Ubiquitous Watergate: An attributional analysis." *Journal of Personality and Social Psychology* 32:66–68.

————. 1976. Ethical issues in the conduct of research in social relations. In *Research methods in social relations,* ed. C. Selltiz, L. S. Wrightsman, and S. W. Cook. New York: Holt.

Cook, T. D.; Bean, R. B.; Calder, B. J.; Frey, R.; Krovetz, M. L.; and Reisman, S. R. 1970. Demand characteristics and three conceptions of the frequently deceived subject. *Journal of Personality and Social Psychology* 14:185–94.

Cooper, J. 1976. Deception and role playing: On telling the good guys from the bad guys. *American Psychologist* 31:605–10.

Coser, L. A. 1959. A question of professional ethics? *American Sociological Review* 24:397–98.

Cox, D. E., and Sipprelle, C. N. 1971. Coercion in participation as a research subject. *American Psychologist* 26:726–31.

Cronbach, L. J. 1970. *Essentials of psychological testing.* New York: Harper.

————. 1977. Remarks to the new society. *Evaluation Research Society Newsletter* 1 (1): 1–3.

Crowne, D. P., and Marlowe, D. 1964. *The approval motive: Studies in evaluative dependence.* New York: Wiley.

Culliton, B. J. 1976. Confidentiality: Court declares researcher can protect sources. *Science* 193:467–69.

Darroch, R. K., and Steiner, I. D. 1970. Role playing: An alternative to laboratory research? *Journal of Personality* 38:302–11.

Davis, F. 1961. Comment on "Initial interaction of newcomers in alcoholics anonymous." *Social Problems* 8:364–65.

————. 1969. Genetics of the discovery of DNA. *Trans-Action* 6:53–56.

Davis, F. B. 1974. *Standards for educational and psychological tests.* Washington, D.C.: American Psychological Association.

Davis, J. R., and Fernald, P. S. 1975. Psychology in action: Laboratory experience versus subject pool. *American Psychologist* 30:523–24.

Davison, G. C., and Stuart, R. B. 1975. Behavior therapy and civil liberties. *American Psychologist* 30:755–63.

Deloria, V. 1969. *Custer died for your sins: An Indian manifesto.* New York: Macmillan.

Denner, B. 1967. *Informers and their influence on the handling of illicit information.* Chicago: Midwestern Psychological Association.

Denzin, N. 1968. On the ethics of disguised observation. *Social Problems* 15:502–6.

References

————. 1973. *The values of social science.* New Brunswick, N.J.: Transaction Books.

De Solla Price, D. 1964. Ethics of scientific publication. *Science* 144: 655-57.

Diener, E.; Matthews, R.; and Smith, R. 1972. Leakage of experimental information to potential future subjects by debriefed subjects. *Journal of Experimental Research in Personality* 6:264-67.

Duce, R. A. 1975. Accurate references. *Science* 187:792.

Dunnette, M. D. 1966. Fads, fashions, and folderol in psychology. *American Psychologist* 21:343-52.

DuShane, G. 1964. An unfortunate event. *Science* 144:945-46.

Edsall, G. 1969. A positive approach to the problem of human experimentation. *Daedalus* 98:463-79.

Edwards, J., and Greenwald, M. 1977. Factors affecting the judged value and ethicality of psychological research. Paper presented at the 85th annual convention of the American Psychological Association, San Francisco.

Ehrlich, H. J. 1969. The impossible possible philosopher's man: A discussion of the sociologist as activist. *Sociological Focus* 2:31-35.

Elms, A. C. 1977. Up front. *Psychology Today* 10:17.

Epstein, C. F. 1975. Ethics committee concerns cited: Reactions invited. *Footnotes* 3:185.

Epstein, Y. M.; Suedfeld, P.; and Silverstein, S. J. 1973. Subjects' expectations of and reactions to some behaviors of experimenters. *American Psychologist* 28:212-21.

Erikson, K. T. 1967. A comment on disguised observation in sociology. *Social Problems* 14:366-73.

Ethical standards for research with children. Society for Research in Child Development. 1973. In R. E. Schutz, Ethical standards for research in education. *Educational Researcher* 2:4-5.

Ethics and the anthropologist. 1976. *Anthropology Newsletter* 17:2.

————. 1977. *Anthropology Newsletter* 18:13.

Evaluation. A journal published by the Program Evaluation Project, Minneapolis Medical Research Foundation, 619 S. Fifth Avenue, Minneapolis, Minn. 55415.

Evaluation Quarterly. Beverly Hills, Calif. Sage Publications.

Evans, P. 1976. The Burt affair ... Sleuthing in science. *APA Monitor* 7:1, 4.

Eysenck, H. 1969. The technology of consent. *New Scientist* 42:688-90.

————. 1977. The case of Sir Cyril Burt: On fraud and prejudice in a scientific controversy. *Encounter* 48:19-24.

Farr, J. L., and Seaver, W. B. 1975. Stress and discomfort in psychological research: Subject perceptions of experimental procedures. *American*

Psychologist 30:770–73.

F.A.S. *Public Interest Report.* 1976a. To whom are public interest scientists responsible? 29 (10): 1–2.

———. 1976b. Sample problems public interest scientists face. 29 (10): 5.

———. Special issue: Animal rights. 30 (8): 1–8.

Feldman, J. J.; Hyman, H.; and Hart, C. W. 1951. Interviewer effects on the quality of survey data. *Public Opinion Quarterly* 15:734–61.

Feldman, M. P., and MacCulloch, M. J. 1971. *Homosexual behavior: Therapy and assessment.* Oxford: Pergamon Press.

Ferguson, L. R. 1976. The competence and freedom of children to make choices regarding participation in biomedical and behavioral research. Paper prepared for the National Commission on the Protection of Human Subjects in Biomedical and Behavioral Research, U.S. Department of Health, Education, and Welfare, Bethesda, Md.

Festinger, L.; Riecken, H.; and Schachter, S. 1956. *When prophecy fails.* Minneapolis: University of Minnesota Press.

Fillenbaum, R. S. 1966. Prior deception and subsequent experimental performance: The faithful subject. *Journal of Personality and Social Psychology* 4:532–37.

Fo, W. S., and O'Donnell, C. R. 1975. The buddy system: Effect of community intervention on delinquent offenses. *Behavior Therapy* 6:522–24.

Forward, J.; Canter, R.; and Kirsch, N. 1976. Role-enactment and deception methodologies: Alternate paradigms. *American Psychologist* 31: 595–604.

Foss, D. C. 1977. *The value controversy in sociology.* San Francisco: Jossey-Bass.

Frankel, L. R. 1976. Restrictions to survey sampling—Legal, practical and ethical. In *Perspectives on attitude assessment: Surveys and their alternatives,* ed. H. W. Sinaiko and L. A. Broedling. Champaign, Ill.: Pendleton Publications.

Frankel, M. S. 1976. Ethical issues associated with experimentation in political science. Paper given at the International Political Science Association Convention, Edinburgh, Scotland, 1976.

Fraser, S. C.; Beaman, A. L.; Diener, E.; and Kelem, R. T. 1977. Two, three, and four heads are better than one: Modification of college performance by peer monitoring. *Journal of Educational Psychology* 69: 109–14.

Freedman, J. L. 1969. Role playing: Psychology by consensus. *Journal of Personality and Social Psychology* 12:107–14.

Freund, P. A. 1967. Is the law ready for human experimentation? *American Psychologist* 22:394–99.

Friedman, N. 1967. *The social nature of psychological research.* New York: Basic Books.

Gaertner, S. and Bickman, L. 1971. Effects of race on the elicitation of helping behavior: The wrong number technique. *Journal of Personality*

References

and Social Psychology 20:218-22.

Galliher, J. F. 1973. The protection of human subjects: A re-examination of the professional code of ethics. *American Sociologist* 8:93-100.

Gallin, B. 1959. A case for intervention in the field. *Human Organization* 18:140-44.

Gallo, P. S.; Smith, S.; and Mumford, S. 1973. Effects of deceiving subjects upon experimental results. *Journal of Social Psychology* 89: 99-107.

Galtung, J. 1975. *Peace: Research, education, action.* Vol. 1. Copenhagen: Christian Eljers.

Gardner, J. M. 1971. Mental retardation research: Where do I send it? *Mental Retardation* 9:12-13.

Garrett, T. M. 1968. *Problems and perspectives in ethics.* New York: Sheed and Ward.

Gay, C. 1973. A man collapsed outside a UW building. Others ignore him. What would you do? *University of Washington Daily,* 30 Nov. 1973, pp. 14-15.

Gearing, F. 1960. *Documentary history of the Fox project.* Chicago: University of Chicago Press.

Gergen, K. J. 1973. The codification of research ethics: Views of a doubting Thomas. *American Psychologist* 28:907-12.

Gillie, O. 1977. Did Sir Cyril Burt fake his research on heritability of intelligence? Part 1. *Phi Delta Kappan* 58:469-71.

Glinski, R. J.; Glinski, B. C.; and Slatin, G. T. 1970. Nonnaivety contamination in conformity experiments: Sources, effects, and implications for control. *Journal of Personality and Social Psychology* 16:478-85.

Goldberg, P. A. 1968. Are women prejudiced against women? *Transaction* 5:28-30.

Golding, S. L., and Lichtenstein, E. 1970. Confession of awareness and prior knowledge of deceptions as a function of interview set and approval motivation. *Journal of Personality and Social Psychology* 14:213-23.

Goldstein, J. H.; Davis, R. W.; and Herman, D. 1975. Escalation of aggression: Experimental studies. *Journal of Personality and Social Psychology* 31:162-70.

Goodfeld, J. 1977. Humanity in science: A perspective and plea. *Science* 198:580-85.

Gouldner, A. 1963. Anti-minotaur: The myth of a value free sociology. In *Sociology on trial,* ed. M. Stein and A. Vidich. Englewood Cliffs, N.J.: Prentice-Hall.

Gray, B. 1975. *Human subjects in medical experimentation.* New York: Wiley.

Gray, D. J. 1968. Value-free sociology: A doctrine of hypocrisy and irresponsibility. *Sociological Quarterly* 9:176-85.

Gray, S. W. 1971. Ethical issues in research in early childhood intervention.

241

Children 18:83–89.

Grayson, D. K. 1969. Human life vs. science. *Anthropology Newsletter* 10 (6):3.

Green, A. 1972. *Sociology: An analysis of life in modern society*. New York: McGraw-Hill.

Greenberg, M. S. 1967. Role playing: An alternative to deception? *Journal of Personality and Social Psychology* 7:152–57.

Griffin, J. H. 1961. *Black like me*. Boston: Houghton-Mifflin.

Gruder, L. L.; Stumpfhauser, A.; and Wyer, R. S. 1977. Improvement in experimental performance as a result of debriefing about deception. *Personality and Social Psychology Bulletin* 3:434–37.

Guest, L. 1947. A study of interviewer competence. *International Journal of Opinion and Attitude Research* 1:17–30.

Guetzkow, H.; Alger, C. F.; Brody, R. A.; Noel, R. C.; and Snyder, R. C. 1963. *Simulation in international relations*. Englewood Cliffs, N.J.: Prentice-Hall.

Guide for the care and use of laboratory animals. 1974. Washington, D.C.: U.S. Government Printing Office.

Guidelines for the use of animals in school science behavior projects. 1972. *American Psychologist* 27:377.

Gustav, A. 1962. Students' attitudes toward compulsory participation in experiments. *Journal of Psychology* 53:119–25.

Guttentag, M., and Struening, E. L., eds. 1975. *Handbook of evaluation research*. Vol. 2. Beverly Hills, Calif.: Sage Publications.

Hamilton, V. L. 1976. Role playing and deception: A re-examination of the controversy. *Journal for the Theory of Social Behavior* 6:233–50.

Hammersla, E. J. 1974. The effects of participation in a laboratory bystander intervention study on subsequent attitudes and intervention behavior. *Dissertation Abstracts International* 35, no. 6 (December): 2989-B.

Hartnett, R. T., and Seligsohn, H. C. 1967. The effects of varying degrees of anonymity on responses to different types of psychological questions. *Journal of Educational Measurement* 4:95–103.

Hathaway, S. R., and McKinley, J. C. 1951. *The Minnesota Multiphasic Personality Inventory*. New York: Psychological Corporation.

Haynes, S. N.; Griffin, P.; Mooney, D.; and Parise, M. 1975. Electromyographic biofeedback and relaxation instructions in the treatment of muscle contraction headaches. *Behavior Therapy* 6:672–78.

Haywood, H. C. 1977. The ethics of doing research . . . and of not doing it. *American Journal of Mental Deficiency* 81:311–17.

Helmreich, W. B. 1973. *Black crusaders: A case study of a black militant organization*. New York: Harper and Row.

Helmstadter, G. C. 1970. *Research concepts in human behavior: Educa-*

References

tion, psychology, sociology. New York: Appleton-Century-Crofts.

Hendrick, C., ed. 1977. Role-playing as a methodology for social research: A symposium. *Personality and Social Psychology Bulletin* 3:454–523.

Henle, M., and Hubbell, M. B. 1938. Egocentricity in adult conversation. *Journal of Social Psychology* 9:227–34.

Hess, E. H. 1965. Attitude and pupil size. *Scientific American* 212:46–54.

Hill, A. B. 1963. Medical ethics and controlled trials. *British Medical Journal* 1:1043–49.

Hirsch, J. 1971. Behavior-genetic analysis and its biosocial consequences. In *Intelligence: Genetic and environmental influences,* ed. R. Cancro. New York: Grune and Stratton.

Hobbs, N. 1959. Science and ethical behavior. *American Psychologist* 14: 217–25.

Holmberg, A. R. 1958. The research and development approach to the study of change. *Human Organization* 17:12–16.

Holmes, D. S. 1976*a*. Debriefing after psychological experiments. I. Effectiveness of postdeception dehoaxing. *American Psychologist* 31:858–67.

———. 1976*b*. Debriefing after psychological experiments. II. Effectiveness of postexperimental desensitizing. *American Psychologist* 31:868–75.

Holmes, D. S., and Bennett, D. H. 1974. Experiments to answer questions raised by the use of deception in psychological research. *Journal of Personality and Social Psychology* 29:358–67.

Hook, S.; Kurtz, P.; and Tôdôrôvich, M., eds. 1977. *Ethics of teaching and scientific research.* Buffalo: Prometheus Books.

Horowitz, I. L., ed. 1967. *The rise and fall of Project Camelot.* Cambridge, Mass.: MIT Press.

Horowitz, I. A., and Rothschild, B. H. 1970. Conformity as a function of deception and role playing. *Journal of Personality and Social Psychology* 14:224–26.

Horvitz, D. G.; Greenberg, B. G.; and Abernathy, J. R. 1976. The randomized response technique. In *Perspectives on attitude assessment: Surveys and their alternatives,* ed. H. W. Sinaiko and L. A. Broedling. Champaign, Ill.: Pendleton Publications.

House, E. R. 1973. The conscience of educational evaluation. *School evaluation: The politics and the process.* Berkeley: McCutchan.

———. 1976. Justice in evaluation. In *Evaluation studies,* vol. 1, ed. G. V. Glass. Beverly Hills, Calif.: Sage Publications.

Hudson, L. 1972. *The cult of the fact.* New York: Harper and Row.

Human Organization. 1958–60. Freedom and responsibility in research: The Springdale case. Vols. 17–19. Editorial and comments follow in next issues of journal.

Humphreys, L. 1970. *Tearoom trade.* Chicago: Aldine.

Humphreys, L. G. 1976. The role of the board of social and ethical re-

243

sponsibility. Unpublished paper. University of Illinois, Champaign, Ill.

Hyman, H. H.; Cobb, W. J.; Feldman, J. J.; Hart, C. W.; and Stember, C. W. 1954. *Interviewing in social research.* Chicago: University of Chicago Press.

Jackson, D. N., and Messick, S. 1967. *Problems in human assessment.* New York: McGraw-Hill.

Jensen, A. R. 1977. Did Sir Cyril Burt fake his research on heritability of intelligence? Part 2. *Phi Delta Kappan* 58:471–92.

Jessor, S. L., and Jessor, R. 1975. Transition from virginity to nonvirginity among youth: A social-psychology study over time. *Developmental Psychology* 11:473–84.

Jones, D. J. 1971. Social responsibilities and the belief in basic research: An example from Thailand. *Current Anthropology* 12:347–50.

Jorgenson, J. 1971. On ethics and anthropology. *Current Anthropology* 12:321–34.

Jourard, S. M. 1968. *Disclosing man to himself.* Princeton, N.J.: Van Nostrand.

Katz, J. 1972. *Experimentation with human beings.* New York: Russell Sage Foundation.

Keith-Spiegel, P. 1976. Children's rights as participants in research. In *Children's rights and the mental health professions,* ed. G. P. Koocher. New York: Wiley.

Kelman, H. C. 1965. Manipulation of human behavior: An ethical dilemma for the social scientist. *Journal of Social Issues* 21:31–46.

———. 1967. Human use of human subjects: The problem of deception in social psychological experiments. *Psychological Bulletin* 67:1–11.

———. 1968. *A time to speak: On human values and social research.* San Francisco: Jossey-Bass.

———. 1978. Research, behavioral. In *Encyclopedia of bioethics,* ed. W. T. Reich. New York: Free Press.

———. In press. Privacy and research with human beings. *Journal of Social Issues.*

Kennedy, E. C. 1975. *Human rights and psychological research.* New York: Thomas Y. Crowell.

Kershaw, D. N. 1972. A negative-income-tax experiment. *Scientific American* 227:19–25.

———. 1975. *The experience with ethical issues in the conduct of social experiments.* Princeton, N.J.: Mathematica.

Kershaw, D. N., and Skidmore, F. 1974. *The New Jersey graduated work incentive experiment.* Princeton, N.J.: Mathematica.

Kimble, G. 1976. The role of risk/benefit analysis in the conduct of psychological research. Paper prepared for the National Commission for the Protection of Human Subjects of Biomedical and Behavioral Research, U.S. Department of Health, Education, and Welfare, Bethesda, Md.

References

King, D. J. 1970. The subject pool. *American Psychologist* 25:1179–81.

Kirkham, G. L. 1975. Doc Cop. *Human Behavior* 4:16–23.

Klaus, R. A., and Gray, S. W. 1968. Early training project for disadvantaged children: A report after five years. *Society for Research in Child Development* vol. 33, no. 4.

Klein, R. E.; Habicht, J. P.; and Yarbrough, C. 1973. Some methodological problems in field studies on nutrition and intelligence. In *Nutrition, development and social behavior,* ed. D. J. Kallen. Washington, D.C.: U.S. Government Printing Office.

Kluckhohn, F. R. 1940. The participant-observer technique in small communities. *American Journal of Sociology* 46:331–43.

Knerr, C. R. 1976. Compulsory disclosure to the courts of research sources and data. Paper presented at the American Psychological Association convention, Washington, D.C.

Koestler, A. 1971. *The case of the midwife toad.* New York: Random House.

Koocher, G. P. 1974. Conversations with children about death—ethical considerations in research. *Journal of Clinical Child Psychology* 3:19–21.

———. 1977. Bathroom behavior and human dignity. *Journal of Personality and Social Psychology* 35:120–21.

Krapfl, J. E., and Vargas, E. A., eds. 1977. *Behavior analysis and ethics.* Kalamazoo, Mich.: Behaviordelia.

Kuhn, T. S. 1970. *The structure of scientific revolutions.* 2d ed. rev. Chicago: University of Chicago Press.

Langer, E. J., and Rodin, J. 1976. The effects of choice and enhanced personal responsibility for the aged: A field experiment in an institutional setting. *Journal of Personality and Social Psychology* 34:191–98.

Lappe, M. 1975. Accountability in science. *Science* 187:696–98.

Latané, B., and Darley, J. M. 1970. *The unresponsive bystander: Why doesn't he help?* New York: Appleton.

Lazarus, R. A. 1964. A laboratory approach to the dynamics of psychological stress. *American Psychologist* 19:400–411.

Lear, J. 1966. Experiments on people—the growing debate. *Saturday Review* 49:41–43.

Levine, R. J. 1975a. The nature and definition of informed consent in various research settings. Paper prepared for the National Commission for the Protection of Human Subjects of Biomedical and Behavioral Research, U.S. Department of Health, Education, and Welfare, Bethesda, Md.

———. 1975b. The role of assessment of risk-benefit criteria in the determination of the appropriateness of research involving human subjects. Paper prepared for the National Commission for the Protection of Human Subjects of Biomedical and Behavioral Research, U.S. Department of Health, Education, and Welfare, Bethesda, Md.

245

Levy, L. H. 1967. Awareness, learning, and the beneficent subject as expert witness, *Journal of Personality and Social Psychology* 6:365-70.

Lewin, K. 1948. *Resolving social conflicts.* New York: Harper.

Lewis, O. 1959. *Five families: Mexican case studies in the culture of poverty.* New York: Wiley.

———. 1963. *The children of Sanchez: Autobiography of a Mexican family.* New York: Vintage Books.

Lichtenstein, E. 1970. Please don't talk to anyone about this experiment: Disclosure of deception by debriefed subjects. *Psychological Reports* 26:485-86.

Lindzey, G. 1950. An experimental examination of the scapegoat theory of prejudice. *Journal of Abnormal and Social Psychology* 45:296-309.

Lofland, J. F., and Lejeune, R. A. 1960. Initial interaction of newcomers in alcoholics anonymous: A field experiment in class symbols and socialization. *Social Problems* 1960 8:102-11.

Lovell, V. R. 1967. The human use of personality tests: A dissenting view. *American Psychologist* 22:383-93.

Lundberg, G. A. 1947. *Can science save us?* New York: Longmans and Green.

McCall, G. J., and Simmons, J. L. 1969. *Issues in participant observation.* Reading, Mass.: Addison-Wesley.

McCartney, J. L. 1970. On being scientific: Changing styles of presentation of sociological research. *American Sociologist* 5:30-35.

———. 1976. Confronting the journal publication crisis: A proposal for a council of social science editors. *American Sociologist* 11:144-52.

McClelland, D. C., and Steele, R. S. 1972. *Motivation workshops.* New York: General Learning Company.

McConaghy, N. 1969. Subjective and penile plethysmograph responses following aversion, relief and apomorphine aversion therapy for homosexual impulses. *British Journal of Psychiatry* 115:723-30.

McGuire, W. J. 1967. Some impending reorientations in social psychology: Some thoughts provoked by Kenneth Ring. *Journal of Experimental Social Psychology* 3:124-39.

———. 1969. Suspiciousness of the experimenter's intent. In *Artifact in behavioral research,* ed. R. Rosenthal and R. L. Rosnow. New York: Academic Press.

MacKinney, A. C. 1955. Deceiving experimental subjects. *American Psychologist* 10:133.

McNemar, Q. 1960. At random: Sense and nonsense. *American Psychologist* 15:295-300.

MacRae, D. 1976. *The social function of social science.* New Haven: Yale University Press.

Mahoney, M. J. 1978. Publish and perish. *Human Behavior* 7:38-41.

References

Martin, D. B. H., and Parsons, C. W. 1973. A code of fair information practice for statistical-reporting and research operations. *Evaluation* 1:31-36.

Masling, J. 1966. Role-behavior of the subject and psychologist and its effects upon psychological data. *Nebraska Symposium on Motivation* 14:67-103.

Masters, W. H., and Johnson, V. E. 1966. *Human sexual response.* Boston: Little, Brown.

Mathes, E. W., and Guest, T. A. 1976. Anonymity and group antisocial behavior. *Journal of Social Psychology* 100:257-62.

Mead, M. 1961. The human study of human beings. *Science* 133:163.

———. 1962. The social responsibility of the anthropologist. *Journal of Higher Education* 33:1-12.

———. 1969. Research with human beings: A model derived from anthropological field practice. *Daedalus* 98:361-86.

Menges, B. 1970. Survey of Psychology 100 students. Unpublished report, Psychology Department, University of Illinois.

Menges, R. J. 1973. Openness and honesty versus coercion and deception in psychological research. *American Psychologist* 28:1030-34.

Merritt, C. B., and Fowler, R. G. 1948. The pecuniary honesty of the public at large. *Journal of Abnormal and Social Psychology* 43:90-93.

Merton, R. K. 1957a. Priorities in scientific discovery: A chapter in the sociology of science. *American Sociological Review* 22: 635-59.

———. 1957b. *Social theory and social structure.* Glencoe, Ill.: Free Press.

Milgram, S. 1963. Behavioral study of obedience. *Journal of Abnormal and Social Psychology* 67:371-78.

———. 1964. Issues in the study of obedience: A reply to Baumrind. *American Psychologist* 19:848-52.

Milgram, S.; Mann, L.; and Harter, S. 1965. The lost letter technique: A tool of social research. *Public Opinion Quarterly* 29:437-38.

Milgram, S., and Shotland, L. 1973. *Television and antisocial behavior: Field experiments.* New York: Academic Press.

Miller, A. 1972. Role playing: An alternative to deception? *American Psychologist* 27:623-36.

Miller, A. R. 1971. *The assault on privacy: Computers, data banks, and dossiers.* Ann Arbor: University of Michigan Press.

Miller, G. A. 1969. Psychology as a means of promoting human welfare. *American Psychologist* 24:1063-75.

Mills, C. W. 1959. *The sociological imagination.* New York: Oxford University Press.

Mills, J. 1976. A procedure for explaining experiments involving deception. *Personality and Social Psychology Bulletin* 2:3-13.

Mittlemist, D.; Knowles, E. S.; and Matter, C. F. 1976. Personal space invasions in the lavatory: Suggestive evidence for arousal. *Journal of Personality and Social Psychology* 33:541-46.

247

———. 1977. What to do and what to report: A reply to Koocher. *Journal of Personality and Social Psychology* 35:122–24.

Mixon, D. 1972. Instead of deception. *Journal for the Theory of Social Behavior* 2:145–77.

———. 1974. If you won't deceive, what can you do? In *Reconstructing social psychology,* ed. N. Armistead. Baltimore: Penguin.

Moreno, J. L. 1953. *Who shall survive?* New York: Beacon House.

Nance, J. 1975. *The gentle Tassaday.* New York: Harcourt Brace Jovanovich.

National Academy of Sciences. 1975. *Experiments and research with humans: Values in conflict.* Washington, D.C.: National Academy of Sciences.

National Commission for the Protection of Human Subjects of Biomedical and Behavioral Research. 1976. Children. Draft proposal on guidelines governing research with children, October 1976.

———. 1977. Protection of human subjects: Research involving prisoners. *U.S. Federal Register,* 14 January, 42 (10): 3076–91 (45CFR, part 46).

———. 1978. Research involving children. *Federal Register,* 13 January, 43(9): 2084–2114.

Nay, W. R. 1975. A systematic comparison of instructional techniques for parents. *Behavior Therapy* 6:14–21.

Nejelski, P. 1976. *Social research in conflict with law and ethics.* Cambridge, Mass.: Ballinger.

Nejelski, P., and Finsterbusch, K. 1973. The prosecutor and the researcher: Present and prospective variations on the Supreme Court's Branzburg decision. *Social Problems* 21:3–21.

Nejelski, P., and Lerman, L. M. 1971. A researcher-subject testimonial privilege: What to do before the subpeona arrives. *Wisconsin Law Review* 1971:1085–1148.

Nelson, S. D. 1975. Nature/nurture revisited II: Social political, and technological implications of biological approaches to human conflict. *Journal of Conflict Resolution* 19:734–61.

Newberry, B. H. 1973. Truth-telling in subjects with information about experiments: Who is being deceived? *Journal of Personality and Social Psychology* 25:369–74.

Nickel, T. 1974. The attribution of intention as a critical factor in the relation between frustration and aggression. *Journal of Personality* 42:482–92.

Novak, E.; Seckman, C. E.; and Stewart, R. D. 1977. Motivations for volunteering as research subjects. *Journal of Clinical Pharmacology* 17:365–71.

Nuremberg Code. 1964. Reprinted in *Science* 143:553.

Oppenheimer, J. R. 1956. Analogy in science. *American Psychologist* 11:127–35.

References

Opton, E. J., Jr. 1974. Psychiatric violence against prisoners when therapy is punishment. *Mississippi Law Journal* 45:605–44.

Orlans, H. 1967. Ethical problems in the relations of research sponsors and investigators: In *Ethics, politics, and social research,* ed. G. Sjoberg. Cambridge, Mass.: Schenkman.

Orne, M. T. 1962. On the social psychology of the psychology experiment: With particular reference to demand characteristics and their implications. *American Psychologist* 17:776–83.

Orne, M. T., and Evans, F. J. 1965. Social control in the psychological experiment: Antisocial behavior and hypnosis. *Journal of Personality and Social Psychology* 1:189–200.

Over, R., and Smallman, S. 1973. Maintenance of individual visibility in publication of collaborative research by psychologists. *American Psychologist* 28:161–66.

Page, E. B. 1958. Teacher comments and student performance: A seventy-four classroom experiment in school motivation. *Journal of Educational Psychology* 49:173–81.

Panati, C., and MacPherson, M. 1976. An epitaph for Sir Cyril? *Newsweek* 88:76.

Paul, G. L. 1966. *Insight vs. desensitization in psychotherapy.* Stanford, Calif.: Stanford University Press.

Perrine, M. W., and Wessman, A. E. 1954. Disguised public opinion interviewing with small samples. *Public Opinion Research* 18:92–96.

Peters, C. B. 1976. Multiple submissions: Why not? *American Sociologist* 11:165–68.

Phillips, H. H. 1976. DHEW regulations governing the protection of human subjects and non-DHEW research: A Berkeley view. Paper prepared for the National Commission for the Protection of Human Subjects of Biomedical and Behavioral Research, U.S. Department of Health, Education, and Welfare, Bethesda, Md.

Piliavin, J., and Piliavin, I. 1972. Effects of blood on reactions to a victim. *Journal of Personality and Social Psychology* 23:353–61.

Platt, J. R. 1964. Strong inference. *Science* 146:347–55.

Polanyi, M. 1964. *Personal knowledge: Towards a post-critical philosophy.* New York: Harper and Row.

Privacy Journal (a journal on privacy in a computer age. P.O. Box 8844, Washington, D.C. 20003).

Privacy Protection Study Commission. 1977. *Personal privacy in an information society.* Washington, D.C.: U.S. Government Printing Office.

Queen, S. A. 1959. No "garrison state"—Difficulties, yes. *American Sociological Review* 24:399–400.

Rainwater, L., and Pittman, D. J. 1967. Ethical problems in studying a politically sensitive and deviant community. *Social Problems* 14:357–66.

Rathje, W. L., and Hughes, W. W. 1976. The garbage project as a non-reactive approach: Garbage in—garbage out. In *Perspectives on attitude assessment: Surveys and their alternatives,* ed. H. W. Sinaiko and L. A. Broedling. Champaign, Ill.: Pendleton Publications.

Rattray, R. S. 1956. *Ashanti law and constitution.* London: Oxford University.

Reiss, A. J. 1968. Police brutality—answers to key questions. *Transaction* 5:10-19.

———. 1971. *The police and the public.* New Haven: Yale University.

———. 1973. Social control of sociological knowing. Paper presented at the American Sociological Association Convention, New York City.

———. 1976. Selected issues in informed consent and confidentiality with special reference to behavioral/social science research/inquiry. National Commission for the Protection of Human Subjects of Biomedical and Behavioral Research, U.S. Department of Health, Education, and Welfare, Bethesda, Md.

Rensberger, B. 1977. Fraud in research is a rising problem in science. *New York Times* 126:1.

Resnick, J. H., and Schwartz, T. 1973. Ethical standards as an independent variable in psychological research. *American Psychologist* 28:134-39.

Reynolds, P. D. 1972. On the protection of human subjects and social science. *International Social Science Journal* 24:693-719.

Riecken, H. W. 1975. Social experimentation. *Society* 12:34-41.

Riecken, H. W., and Boruch, R. F. 1974. *Social experimentation: A method for planning and evaluating social intervention.* New York: Academic Press.

Riegel, K. F. 1972. Influence of economic and political ideologies on the development of developmental psychology. *Psychological Bulletin* 78: 129-41.

Riesman, D., and Watson, J. 1964. The sociability project: A chronicle of frustration and achievement. In *Sociologists at work,* ed. P. E. Hammond. New York: Basic Books.

Rights of children as research subjects. 1976. Proceedings of conference held at the University of Illinois at Urbana-Champaign, 14-16 October.

Ring, K. 1967. Experimental social psychology: Some sober questions about some frivolous values. *Journal of Experimental Social Psychology* 3:113-23.

Ring, K.; Wallston, K.; and Corey, M. 1970. Mode of debriefing as a factor affecting subjective reaction to a Milgram-type obedience experiment: An ethical inquiry. *Journal of Representative Research in Social Psychology* 1:67-88.

Rivlin, A. M., and Timpane, P. M., eds. 1975. *Ethical and legal issues of social experimentation.* Washington, D.C.: Brookings Institution.

Robinson, J., and Shaver, P. 1973. *Measures of social psychological at-*

References

titudes, rev. ed. Ann Arbor, Mich.: Institute for Social Research.

Rodman, H. 1970. The moral responsibility of journal editors and referees. *American Sociologist* 5:351-57.

Rokeach, M. 1973. *The nature of human values.* New York: Free Press.

Rokeach, M.; Zemach, R.; and Norrell, G. 1966. The pledge to secrecy: A method to assess violations. Unpublished paper, Michigan State University, 1966.

Rosenhan, D. L. 1973. On being sane in insane places. *Science* 179: 250-58.

Rosenthal, R. 1966. *Experimenter effects in behavioral research.* New York: Appleton-Century-Crofts.

Rosenthal, R., and Lawson, R. 1964. A longitudinal study of the effects of experimenter bias on the operant learning of laboratory rats. *Journal of Psychiatric Research* 2:61-72.

Rosenthal, R., and Rosnow, R. L., eds. 1969. *Artifact in behavioral research.* New York: Academic Press.

Ross, L.; Lepper, M.; and Hubbard, M. 1975. Perseverence in self-perception and social perception: Biased attributional processes in the debriefing paradigm. *Journal of Personality and Social Psychology* 32: 880-92.

Roth, J. 1962. Comments on secret observation. *Social Problems* 9:283-84.

————. 1966. Hired hand research. *American Sociologist* 1:190-96.

Rubin, Z. 1970. Jokers wild in the lab. *Psychology Today* 4:18-24.

Ruebhausen, O. M., and Brim, O. G., Jr. 1966. Privacy and behavioral research. *American Psychologist* 21:423-44.

Rugg, E. A. 1975. Ethical judgments of social research involving experimental deception. *Dissertation Abstracts International* 36:(4-B): 1976.

Ryan, W. 1971. *Blaming the victim.* New York: Pantheon.

Rynkiewich, M. A., and Spradley, J. P. 1976. *Ethics and anthropology: Dilemmas in fieldwork.* New York: Wiley.

Sanford, N. 1965. Will psychologists study human problems? *American Psychologist* 20:192-202.

Schachter, S., and Singer, J. 1962. Cognitive, social and physiological determinants of emotional state. *Psychological Review* 69:379-99.

Schlenker, B. R., and Forsyth, D. R. 1977. On the ethics of psychological research. *Journal of Experimental Social Psychology* 13:369-96.

Schuessler, K. 1976. Reply to Calvin Peters. *American Sociologist* 11:172-73.

Schultz, D. P. 1969. The human subject in psychological research. *Psychological Bulletin* 72:214-28.

Schwitzgebel, R. 1968. Ethical problems in experimentation with offenders. *American Journal of Orthopsychiatry* 38:738-48.

Science and politics: AMA attacked for use of disputed survey in "Medicare" lobbying. 1960. *Science* 132:604-5.

Scriven, M. 1976. Evaluation bias and its control. In *Evaluation studies,* vol. 1, ed. G. V. Glass. Beverly Hills, Calif.: Sage Publications.

Seashore, S. E. 1976. Issues in credit assignment, plagiarism, and ownership of data. Paper presented at the American Psychological Association, Washington, D.C.

Seeman, J. 1969. Deception in psychological research. *American Psychologist* 24:1025-28.

Seidman, E. 1977. Justice, values and social science: Unexamined premises. In *Research in law and sociology,* vol. 1, ed. R. J. Simon. Greenwich, Conn.: Johnson Associates.

Seidman, E.; Rappaport, J.: and Davidson, W. S. 1976. Adolescents in legal jeopardy: Initial success and replication of an alternative to the criminal justice system. Paper presented at the American Psychological Association convention, Washington, D.C.

Sharp, L. 1952. Steel axes for stoneage Australians. *Human Organization* 11:17-22.

Sheatsley, P. B. 1947. Some uses of interviewer report forms. *Public Opinion Quarterly* 11:601-11.

Shils, E. A. 1959. Social inquiry and the autonomy of the individual. In *The human meaning of the social sciences,* ed. D. Lerner. New York: Meridian.

Silverman, I. 1975. Nonreactive methods and the law. *American Psychologist* 30:764-69.

Silverman, I.; Shulman, A. D.; and Wiesenthal, D. L. 1970. Effects of deceiving and debriefing psychological subjects on performance in later experiments. *Journal of Personality and Social Psychology* 14:203-12.

Silvert, K. H. 1965. American academic ethics and social research abroad: The lesson of Project Camelot. *Background* 9:215-36.

Simons, C. W., and Piliavin, J. A. 1972. Effects of deception on reactions to a victim. *Journal of Personality and Social Psychology* 21:56-60.

Sjoberg, G. 1971. *Ethics, politics and social research.* Cambridge, Mass.: Schenkman Publishing.

———. 1975. Politics, ethics and evaluation research. In *Handbook of evaluation research,* vol. 2, ed. M. Guttentag and E. L. Struening. Beverly Hills, Calif.: Sage Publications.

Skinner, B. F. 1971. *Beyond freedom and dignity.* New York: Knopf.

Smith, M. B. 1967. Conflicting values affecting behavioral research with children. *American Psychologist* 22:377-82.

———. 1969. *Social psychology and human values.* Chicago: Aldine.

———. 1973. Is psychology relevant to new priorities? *American Psychologist* 28:463-71.

———. 1974. *Humanizing social psychology.* San Francisco: Jossey Bass.

———. 1976. Some perspectives on ethical/political issues in social research. *Personality and Social Psychology Bulletin* 2:445-53.

Smith, R. E.; Wheeler, G.; and Diener, E. 1975. Faith without works:

References

Jesus people, resistance to temptation and altruism. *Journal of Applied Social Psychology* 5:320–30.

Smith, R. J. 1977*a*. Electroshock experiment at Albany violates ethical guidelines. *Science* 198:383–86.

———. 1977*b*. SUNY at Albany admits research violations. *Science* 198: 708.

Sperry, R. W. 1977. Bridging science and values. A unifying view of mind and brain. *American Psychologist* 32:237–45.

Spiegel, D., and Keith-Spiegel, P. 1970. Assignment of publication credits: Ethics and practices of psychologists. *American Psychologist* 25:738–47.

Spillius, J. 1957. Natural disaster and political crisis in a Polynesian society: An exploration of operational research. *Human Relations* 10:113–25.

Stang, D. J. 1976. Ineffective deception in conformity research: Some causes and consequences. *European Journal of Social Psychology* 6:353–67.

———. Unpublished. An analysis of the deception–role-playing issue.

Steiner, I. D. 1972. The evils of research: Or what my mother didn't tell me about the sins of academia. *American Psychologist* 27:766–68.

Stricker, L. J. 1967. The true deceiver. *Psychological Bulletin* 68:13–20.

Stricker, L. J., Messick, S., and Jackson, D. N. 1967. Suspicion of deception: Implications for conformity research. *Journal of Personality and Social Psychology* 5:379–89.

———. 1969. Evaluating deception in psychological research. *Psychological Bulletin* 71:343–51.

Struening, E. L., and Guttentag, M., eds. 1975. *Handbook of evaluation research*. Vol. 1. Beverly Hills, Calif.: Sage Publications.

Suchman, E. A. 1967. *Evaluation research*. New York: Russell Sage Foundation.

Sullivan, D. S., and Deiker, T. A. 1973. Subject-experimenter perceptions of ethical issues in human research. *American Psychologist* 28:587–91.

Sullivan, M. A.; Queen, S. A.; and Patrick, R. C. 1958. Participant observation as employed in the study of a military training program. *American Sociological Review* 23:660–67.

Swingle, P. G. 1973. *Social psychology in natural settings*. Chicago: Aldine.

Szasz, T. S. 1967. Moral man: A model of man for humanistic psychology. In *Challenges of humanistic psychology,* ed. J. T. Bugental. New York: McGraw-Hill.

Tapp, J. L.; Kelman, H. C.; Triandis, H. C.; Wrightsman, L. S.; and Coelho, G. V. 1974. Continuing concerns in cross-cultural ethics: A report. *International Journal of Psychology* 9:231–49.

Tesch, F. E. 1977. Debriefing research participants: Though this be method there is madness to it. *Journal of Personality and Social Psychology* 35:217–24.

Teuber, H. L. 1960. Perception. In *Handbook of physiology, neuro-*

physiology, vol. 3, ed. J. Fields, H. W. Mogun, and V. E. Hall. Washington, D.C.: American Physiological Association.

Tillery, D. 1967. Seeking a balance between the right of privacy and the advancement of social research. *Journal of Educational Measurement* 4:11.

Tuckman, H. P., and Leaky, J. 1975. What is an article worth? *Journal of Political Economy* 83:951-67.

United States Department of Health, Education, and Welfare. 1971. *The institutional guide to DHEW policy on protection of human subjects.* Washington, D.C.: U.S. Government Printing Office.

Vallance, T. 1972. Social science and social policy: Amoral methodology in a matrix of values. *American Psychologist* 27:107-13.

Vargas, E. A. 1975. Rights: A behavioristic analysis. *Behaviorism* 3:178-90.

———. 1977. Is freedom necessary for rights? In *Behavior analysis and ethics,* ed. J. E. Krapfl and E. A. Vargas. Kalamazoo, Mich.: Behaviordelia.

Vargus, B. S. 1971. On sociological exploitation: Why the guinea pig sometimes bites. *Social Problems* 19:238-48.

Veatch, R. M. 1976. Three theories of informed consent: Philosophical foundations and policy implications. Paper prepared for the National Commission on the Protection of Human Subjects in Biomedical and Behavioral Research, U.S. Department of Health, Education, and Welfare, Bethesda, Md.

Veno, A., and Peeke, H. J. S. 1974. Research on crowding in prisons: Methodological problems and ethical concerns. *Bulletin of the Psychonomic Society* 3:183-84.

Verplanck, W. S. 1955. The control of the content of conversation: Reinforcement of statements of opinion. *Journal of Abnormal and Social Psychology* 55:668-76.

Vidich, A. J. 1960. Freedom and responsibility in research: A rejoinder. *Human Organization* 19:3-4.

Vidich, A. J., and Bensman, J. 1960. *Small town in mass society.* Garden City, N.Y.: Doubleday.

Vinacke, W. E. 1954. Deceiving experimental subjects. *American Psychologist* 9:155.

Wade, N. 1976. I.Q. and heredity: Suspicion of fraud beclouds classic experiment. *Science* 194:916-18.

———. 1977. Thomas S. Kuhn: Revolutionary theorist of science. *Science* 197:143-45.

Wallwork, E. 1975. In defense of substantive rights: A reply to Baumrind. In *Human rights and psychological research*, ed. E. C. Kennedy. New York: Thomas Y. Crowell.

Walsh, J. 1969. ACE study on campus unrest: Questions for behavioral scientists. *Science* 165:151-60.

References

Walsh, W. B., and Stillman, S. M. 1974. Disclosure of deception by debriefed subjects. *Journal of Counseling Psychology* 21:315-19.

Walster, E. 1965. The effect of self-esteem on romantic liking. *Journal of Experimental Social Psychology* 1:184-97.

Walster, E.; Berscheid, E.; Abrahams, D.; and Aronson, V. 1967. Effectiveness of debriefing following deception experiments. *Journal of Personality and Social Psychology* 6:371-80.

Warner, S. L. 1965. Randomized response: A survey technique for eliminating evasive answer bias. *Journal of the American Statistical Association* 60:63-69.

Warwick, D. P. 1975. Deceptive research: Social scientists ought to stop lying. *Psychology Today* 10:38-40.

Warwick, D. P., and Kelman, H. C. 1973. Ethical issues in social intervention. In *Process and phenomena of social change*, ed. G. Zaltman. New York: John Wiley.

Watson, J. 1968. *The double helix.* New York: Atheneum.

Weaver, T. 1973. *To see ourselves: Anthropology and modern social issues.* Glenview, Ill.: Scott, Foresman.

Webb, E.; Campbell, D.; Schwartz, R.; and Sechrest, L. 1966. *Unobtrusive measures: Nonreactive research in the social sciences.* Chicago: Rand McNally.

Webber, M. W. 1973. The politics of information. In *The values of social science*, ed. N. K. Denzin. New Brunswick, N.J.: Transaction Books.

Weinberger, C. W. 1974. Protection of human subjects. *Federal Register,* 30 May 39 (105): 18914-20 (45CFR, part 46).

—————. 1975. Protection of human subjects: Technical amendments. *Federal Register* 13 March 40 (50): 11854-58 (45CFR, part 46).

West, F. E. 1973. *Science for society: A bibliography.* Washington, D.C.: American Association for the Advancement of Science.

West, S. G., and Gunn, S. P. 1978. Some issues of ethics and social psychology. *American Psychologist* 33:30-38.

West, S. G.; Gunn, S. P.; and Chernicky, P. 1975. Ubiquitous Watergate: An attributional analysis. *Journal of Personality and Social Psychology* 32:55-65.

Westin, A. F. 1967. *Privacy and freedom.* New York: Atheneum.

Wilkie, J. R., and Allen, I. L. 1975. Women sociologists and co-authorship with men. *American Sociologist* 10:19-24.

Willis, R. H., and Willis, Y. A. 1970. Role playing versus deception: An experimental comparison. *Journal of Personality and Social Psychology* 16:472-77.

Wilson, D. W., and Donnerstein, E. 1976. Legal and ethical aspects of nonreactive social psychological research: An excursion into the public mind. *American Psychologist* 31:765-73.

255

Wolf, E., and Jorgensen, J. 1970. Anthropology on the warpath in Thailand. *New York Review of Books* 15 (9): 26-35.

Wolfensberger, W. 1967. Ethical issues in research with human subjects. *Science* 155:48-59.

Wolins, L. 1962. Responsibility for raw data. *American Psychologist* 17: 657-58.

Wortman, C. B., Hendricks, M., and Hillis, J. W. 1976. Factors affecting participant reactions to random assignment in social programs. *Journal of Personality and Social Psychology* 33:256-66.

Wuebben, P. L. 1967. Honesty of subjects and birth order. *Journal of Personality and Social Psychology* 5:350-52.

———. 1974. Dissemination of experimental information by debriefed subjects: What is told to whom, when. In *The experiment as a social occasion,* ed. P. L. Wuebben, B. C. Straits, and G. I. Schulman. Berkeley: Glendessary Press.

Wyatt, D. F., and Campbell, D. T. 1950. A study of interviewer bias as related to interviewer's expectations and own opinions. *International Journal of Opinion and Attitude Research* 4:77-83.

Zimbardo, P. G. 1973. On the ethics of intervention in human psychological research: With special reference to the Stanford prison study. *Cognition* 2:243-56.

Zimbardo, P. G.; Haney, C.; Banks, W. C.; and Jaffe, D. 1973. The mind is a formidable jailer: A pirandellian prison. *New York Times Magazine* 122: 8 April, section 6.

Zuckerman, H. A. 1968. Patterns of name ordering among authors of scientific papers: A study of social symbolism and its ambiguity. *American Journal of Sociology* 74:276-91.

Author Index

257

Index

259

Subject Index

Ohio Dominican College Library
1216 Sunbury Road
Columbus, Ohio 43219

GAYLORD